Standard Grade | General | Credit

Geography

General Level 2004
___

Credit Level 2004
___

General Level 2005
___

Credit Level 2005
___

General Level 2006
___

Credit Level 2006
___

General Level 2007
___

Credit Level 2007
___

General Level 2008
___

Credit Level 2008
___

Leckie ⨉ Leckie

© Scottish Qualifications Authority
All rights reserved. Copying prohibited. No part of this publication may be reproduced, stored in a retrieval system, or
transmitted in any form or by any means, electronic, mechanical, photocopying, recording or otherwise.

First exam published in 2004.
Published by Leckie & Leckie Ltd, 3rd Floor, 4 Queen Street, Edinburgh EH2 1JE
tel: 0131 220 6831 fax: 0131 225 9987 enquiries@leckieandleckie.co.uk www.leckieandleckie.co.uk

ISBN 978-1-84372-632-6

A CIP Catalogue record for this book is available from the British Library.

Leckie & Leckie is a division of Huveaux plc.

Leckie & Leckie is grateful to the copyright holders, as credited at the back of the book, for permission to use their material.
Every effort has been made to trace the copyright holders and to obtain their permission for the use of copyright material.
Leckie & Leckie will gladly receive information enabling them to rectify any error or omission in subsequent editions.

[BLANK PAGE]

FOR OFFICIAL USE

G

KU	ES

Total Marks

1260/403

NATIONAL
QUALIFICATIONS
2004

MONDAY, 17 MAY
10.25 AM–11.50 AM

GEOGRAPHY
STANDARD GRADE
General Level

Fill in these boxes and read what is printed below.

Full name of centre

Town

Forename(s)

Surname

Date of birth
Day Month Year   Scottish candidate number   Number of seat

1 Read the whole of each question carefully before you answer it.

2 Write in the spaces provided.

3 Where boxes like this ☐ are provided, put a tick ✓ in the box beside the answer you think is correct.

4 Try all the questions.

5 Do not give up the first time you get stuck: you may be able to answer later questions.

6 Extra paper may be obtained from the invigilator, if required.

7 Before leaving the examination room you must give this book to the invigilator. If you do not, you may lose all the marks for this paper.

SCOTTISH
QUALIFICATIONS
AUTHORITY

Ordnance Survey®

Four colours should appear above; if no
Four colours should appear above; if no

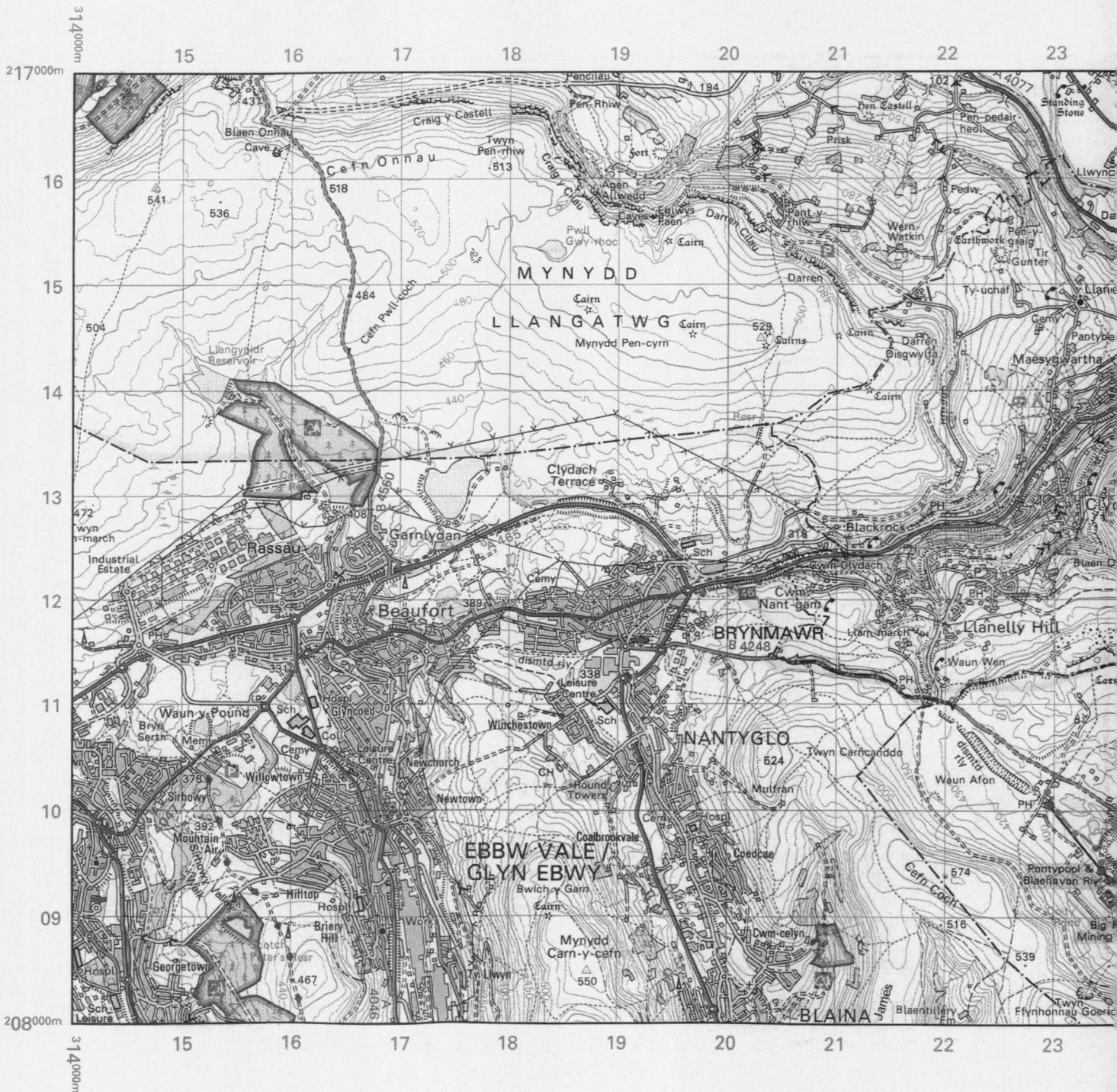

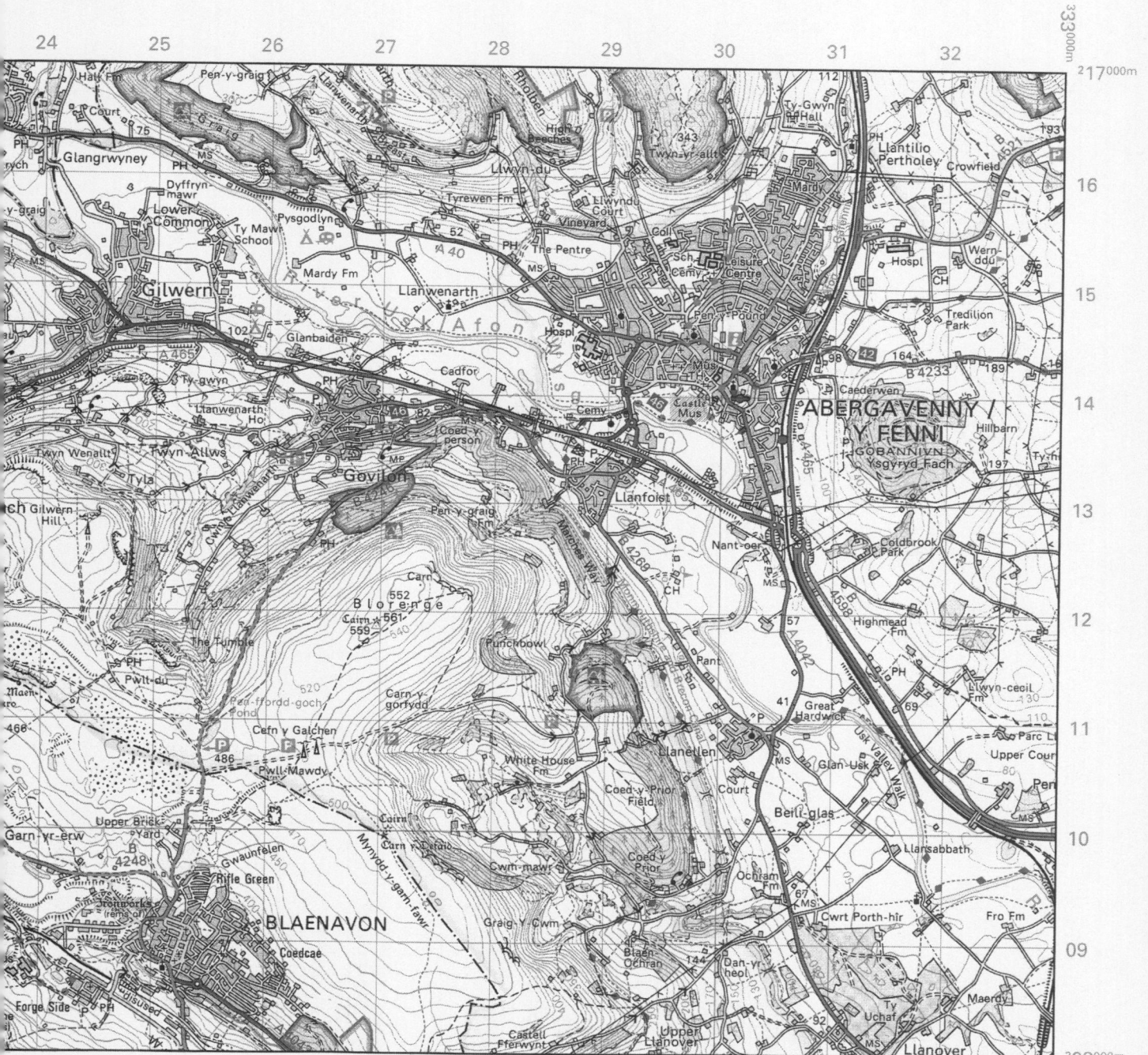

Extract No 1348/161

1:50 000 Scale
Landranger Series

Scale 1: 50 000

2 centimetres to 1 kilometre (one grid square)

Kilometres

Miles

1 kilometre = 0·6214 mile

1 mile = 1·6093 kilometres

1.

Reference Diagram Q1A

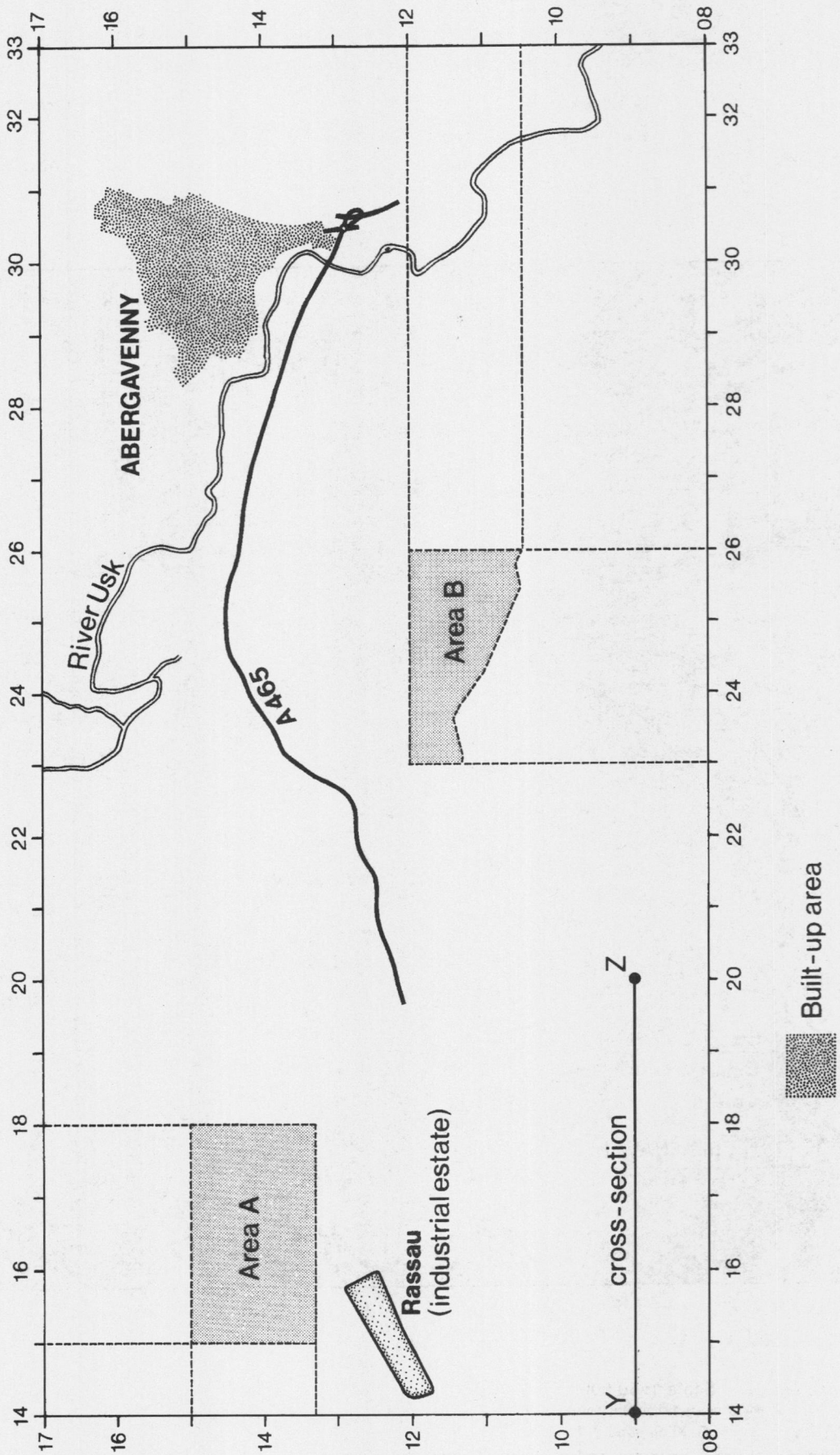

ABERGAVENNY

River Usk

A 465

Area A

Area B

Rassau
(industrial estate)

Y

Z

cross-section

Built-up area

KU	ES

Marks

1. (continued)

Look at the Ordnance Survey Map Extract (No 1348/161) of the Ebbw Vale/Abergavenny area and Reference Diagram Q1A on *Page two*.

(*a*) Complete the table below by matching the physical features to the correct grid references.

Choose from the following grid squares.

1814   2316   2112   1808

Physical Feature	Grid Square
Deep narrow valley	1808
Ridge between two valleys, over 500 metres	2112
Broad flood plain	2316
Part of gentle slope, facing south west	1814

3

[Turn over

Marks

1. (continued)

Reference Diagram Q1B: Cross-section YZ from 140090 to 200090

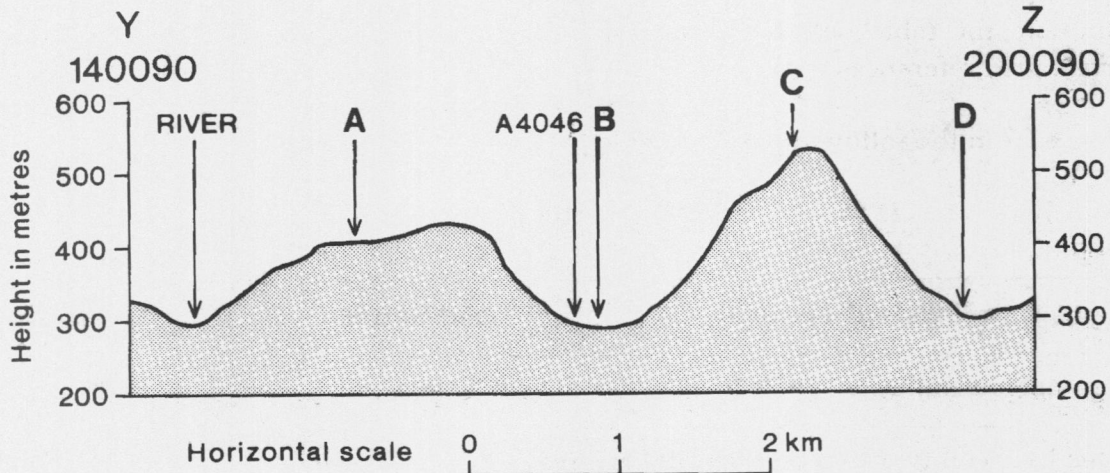

(b) Look at Reference Diagram Q1B. Find this cross-section on the Ordnance Survey map.

Match the features (A, B, C and D) on the cross-section YZ with the correct descriptions in the table below.

Feature	Letter
Works	B
A467	D
Cairn	C
Scotch Peter's Reservoir	A

3

(c) In what ways has the **physical** landscape created problems for engineers in building the A465 road from grid square 1912 to grid square 3012?

Land too steep to build on, lots of cuts made into steep slopes, have to cross a river, embankments need created, national park in way, for low land

4

DO NOT
WRITE IN
THIS
MARGIN

KU | ES

Marks

1. (continued)

(*d*) What is the main function of the town of Abergavenny?

Tick (✓) your choice.

Tourist resort [✓]   Market town [ ]

Give reasons for your choice.

Campsites sarround town, golf course near to to attract tourists, castle museum to attract people. Good place to stay is want to visit National parks

4

(*e*) **Reference Diagram Q1C: Selected Aims of National Parks**

> *   preserve the beauty of the countryside
>
> *   conserve the local wildlife
>
> *   provide good access and facilities for public open air enjoyment
>
> *   maintain established farming

Areas A and B on Reference Diagram Q1A are in the Brecon Beacons National Park. Find them on the map extract.

For each area, **explain** how land use is in conflict with the aims shown above.

Area A _____

_____

_____

Area B _____

_____

_____

4

[Turn over

DO NOT WRITE IN THIS MARGIN

KU	ES

Marks

1. (continued)

(*f*) There is an industrial estate at Rassau, grid squares 1412/1512.

What are the advantages **and** disadvantages of this location for an industrial estate?

Advantages ______________________________

____

____

____

Disadvantages ______________________________

____

____

______________________________ **4**

Marks

2.

Reference Diagram Q2: How a Waterfall develops

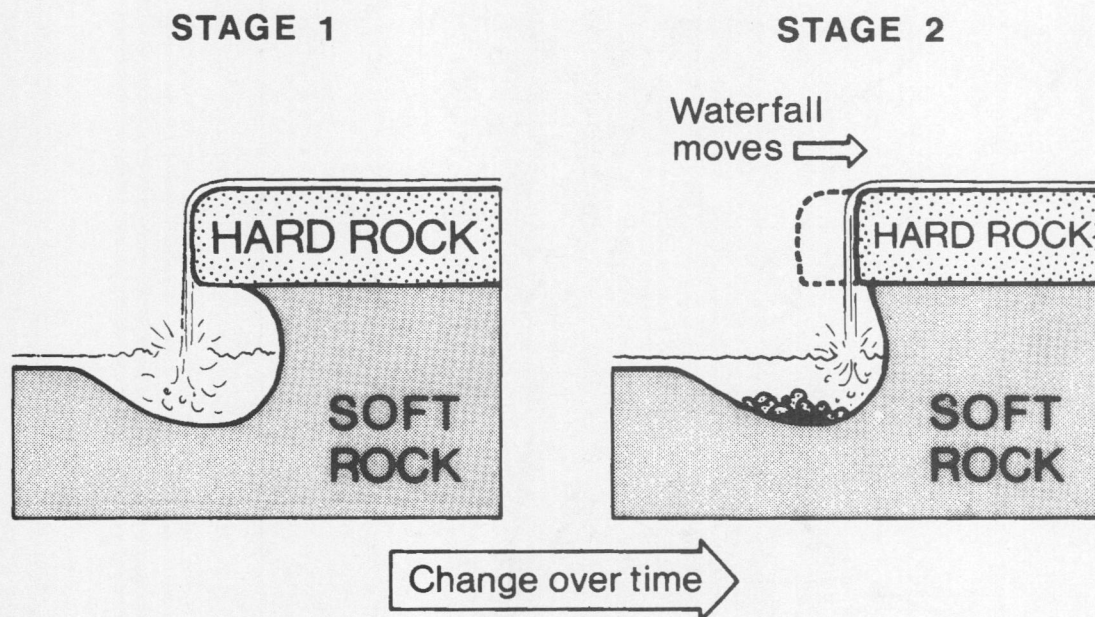

STAGE 1

STAGE 2

Waterfall
moves ⇨

HARD ROCK

HARD ROCK

SOFT
ROCK

SOFT
ROCK

Change over time ⇨

Study Reference Diagram Q2.

Explain, in detail, why a waterfall moves upstream from its original position.

The hard rock on to erodes slowly and gets pushed back, the soft rock erodes faster than the hard rock meaning the hard rock will fall into plunge pool. This happens over may many years.

3

[Turn over

Page seven

DO NOT
WRITE IN
THIS
MARGIN

KU | ES

Marks

3. **Reference Diagram Q3: Synoptic Chart, 12 December 0600 hours**

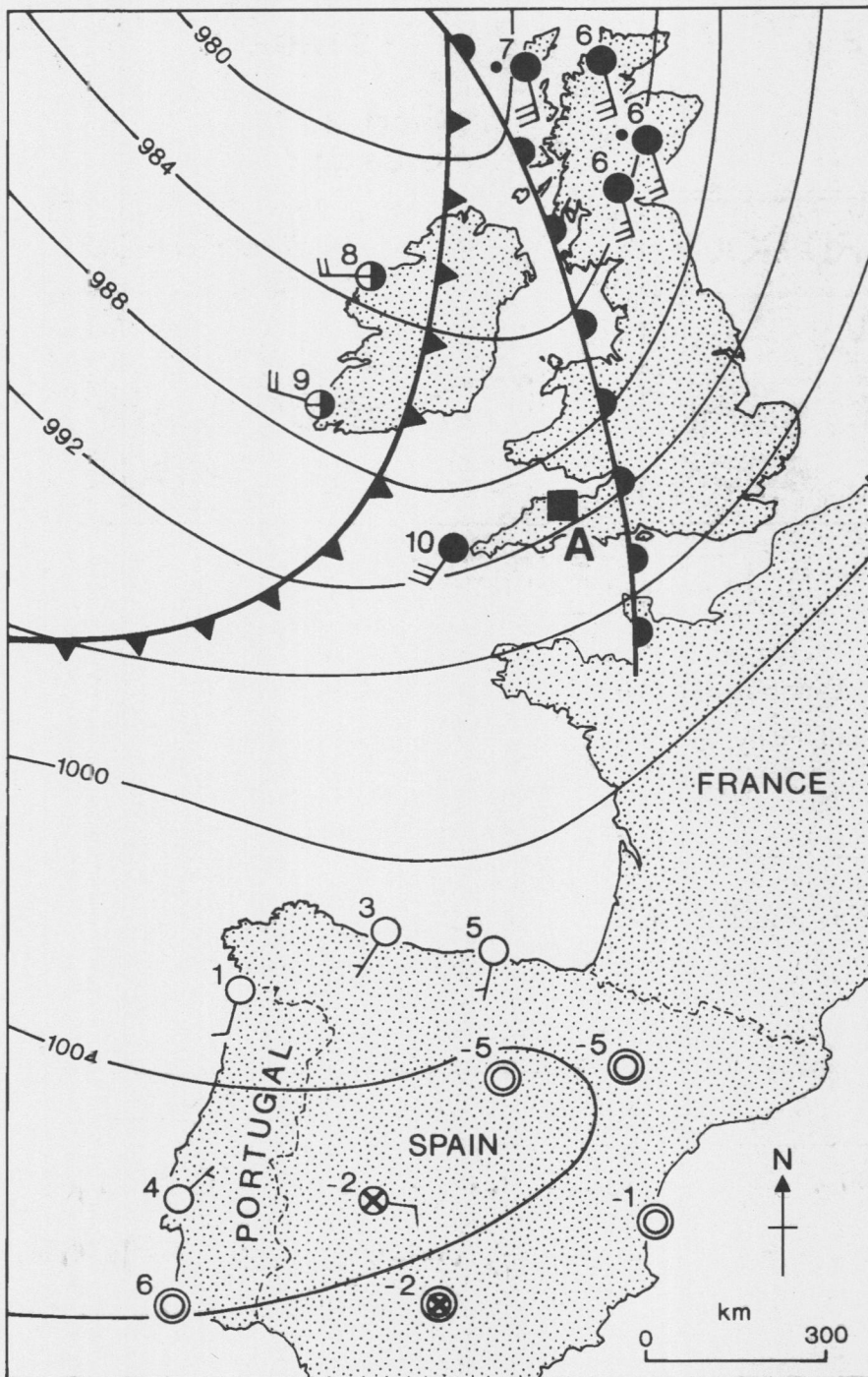

3. (continued)

(*a*) Complete the station circle below to show the weather conditions at **A** on Reference Diagram Q3.

Weather conditions at A

Wind from South West

Cloud cover: 7 oktas

Rain

Wind speed: 15 knots

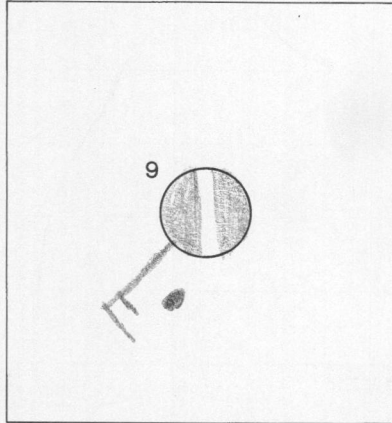

Marks

3

(*b*) Study Reference Diagram Q3.

Match the weather systems to the locations given in the table.

Choose from: Anticyclone   Depression

Location	Weather System
British Isles	Depression
Spain	Anti Cyclone

Give reasons for your answer.

Britain: isobar values lower 1000, fronts appear. Spain: isobars more than 1000, bars are spaced far apart.

4

[Turn over

DO NOT
WRITE IN
THIS
MARGIN

KU | ES

Marks

4.   **Reference Diagram Q4A: Tropical Rainforest Climate**

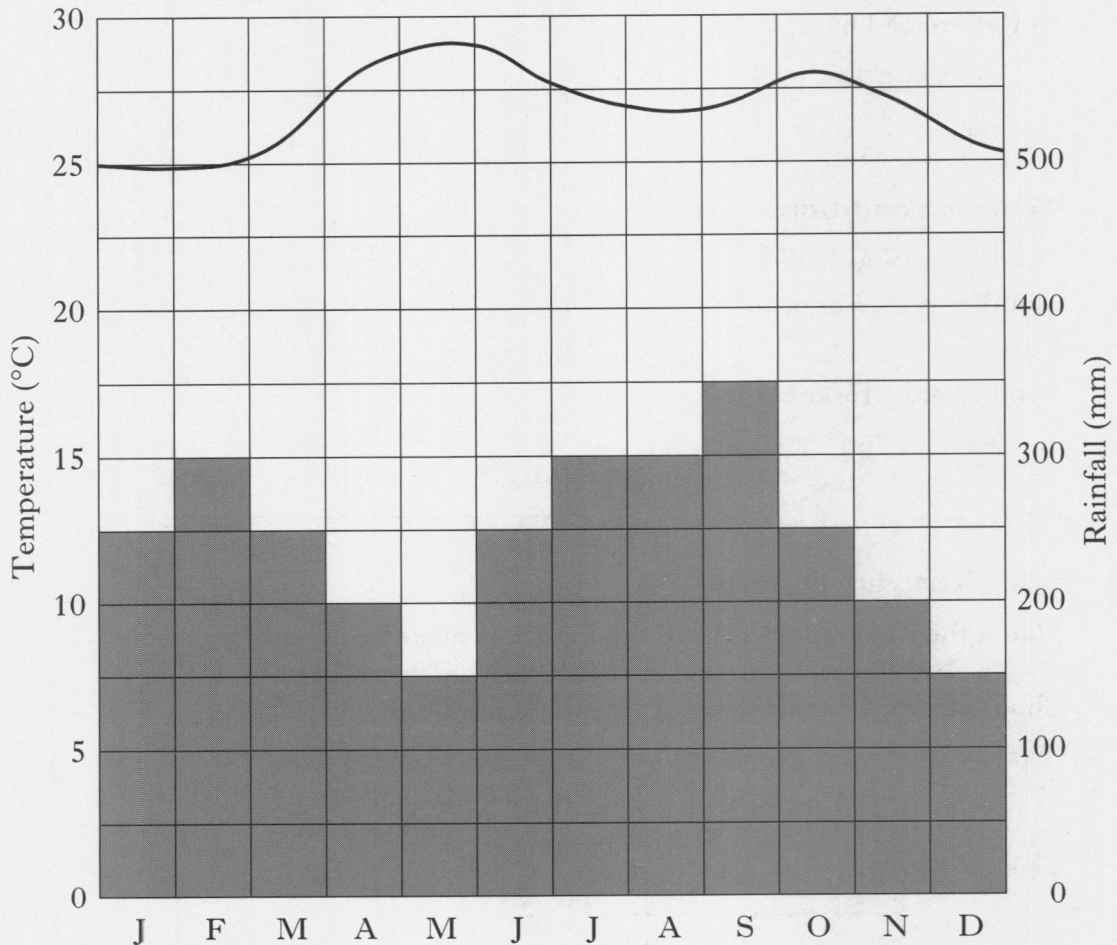

(*a*)   Look at Reference Diagram Q4A.

Describe, **in detail**, the climate of the tropical rainforest.

Temperature constantly high from 25°
to 30°. Constant heavy rainfall
peaking in February and September
reaching 350mm in one month

3

KU	ES

Marks

4. (continued)

**Reference Diagram Q4B: Tropical Rainforest Landscape Before
Deforestation**

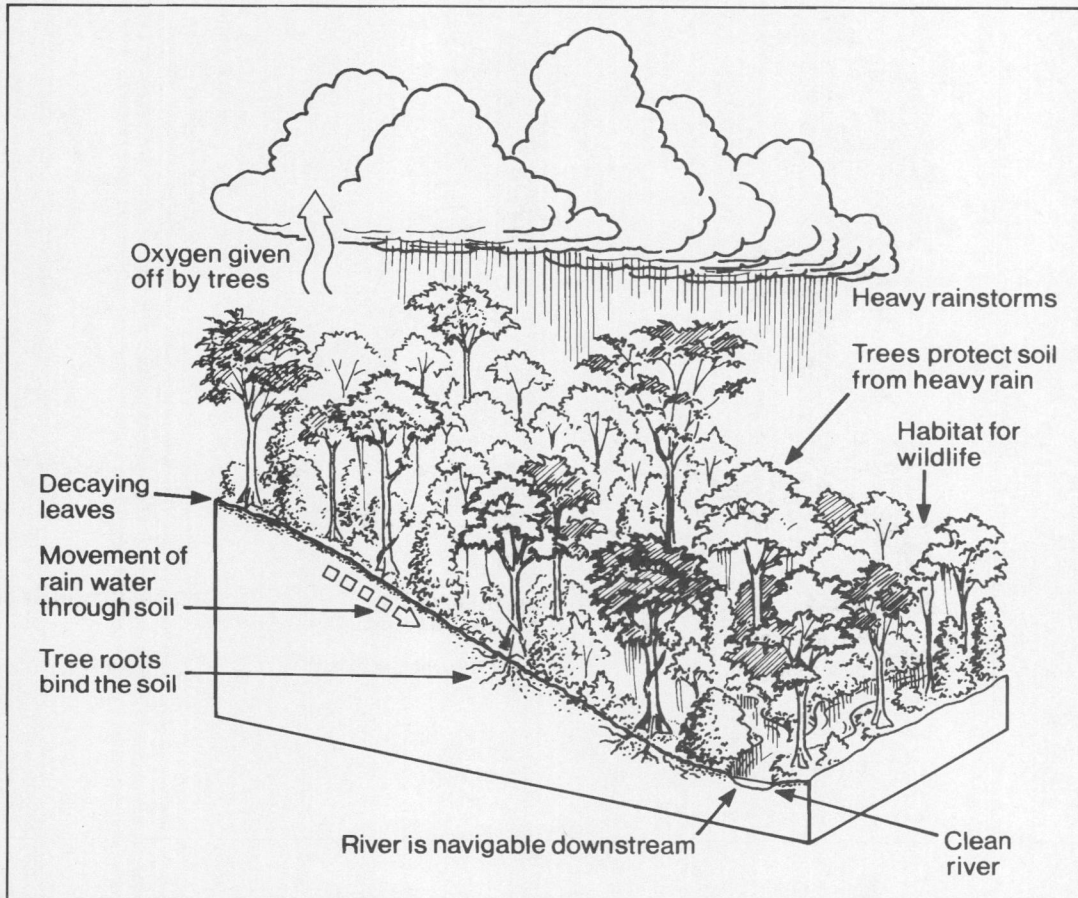

Oxygen given off by trees

Heavy rainstorms

Trees protect soil from heavy rain

Habitat for wildlife

Decaying leaves

Movement of rain water through soil

Tree roots bind the soil

River is navigable downstream

Clean river

(b)   Look at Reference Diagram Q4B.

Explain problems caused by deforestation in the Tropical Rainforest.

Wood required for large population

4

[Turn over

5.   Reference Diagram Q5: Relief, Climate and Selected Land Uses

RELIEF MAP

Land above 200 m

Land below 200 m

CLIMATE MAP

Precipitation

Over 1000 mm

750 - 1000 mm

Under 750 mm

14°C —— 14°C July isotherm

DISTRIBUTION OF ROUGH GRAZING

Rough grazing

DISTRIBUTION OF BARLEY

Barley

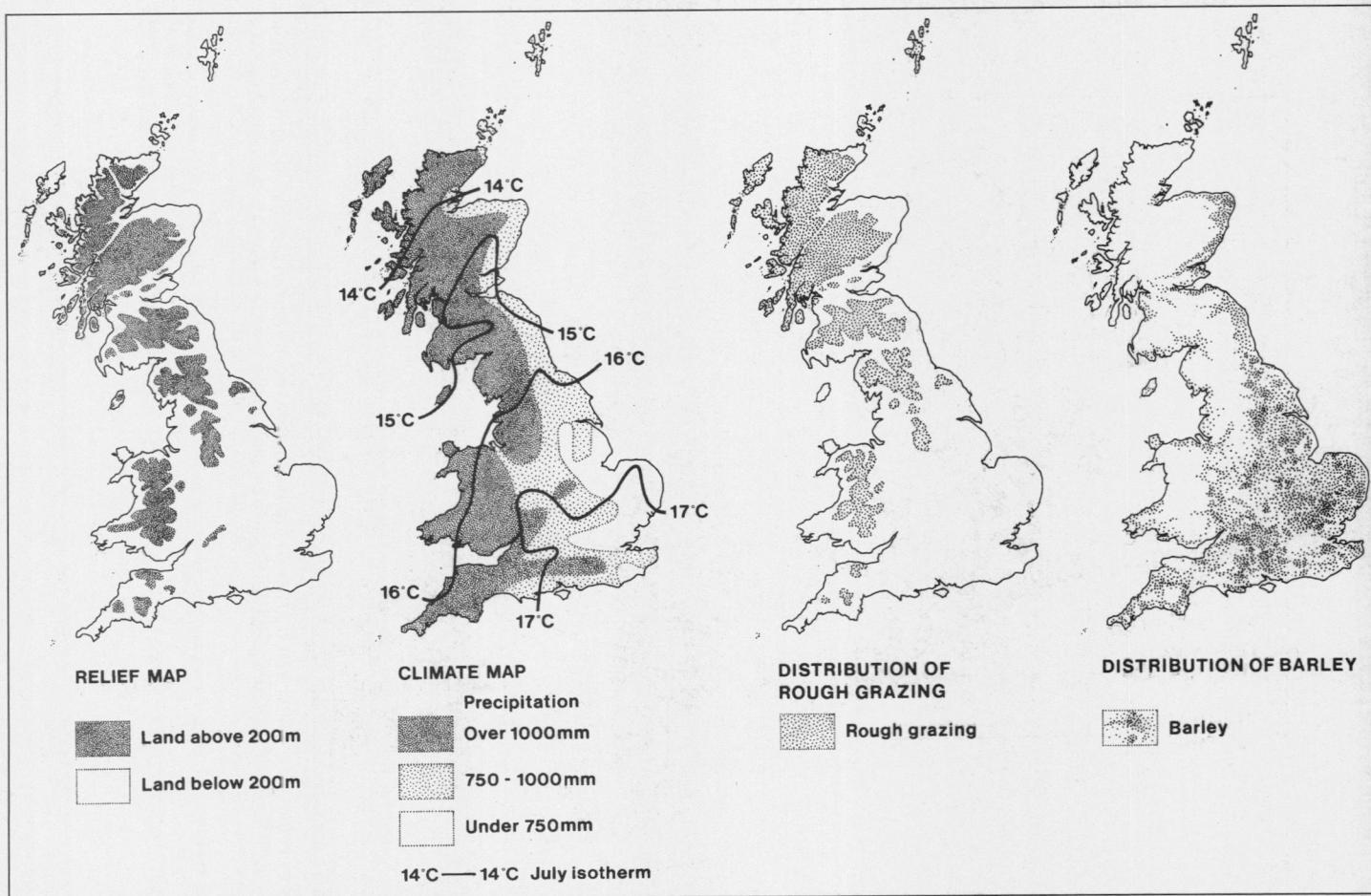

Look at Reference Diagram Q5.

What influence do relief and climate have on the distribution of rough grazing and barley?

DO NOT WRITE IN THIS MARGIN

KU | ES

Marks

4

[Turn over for Question 6 on *Page fourteen*

6.

Reference Diagram Q6A: The Inverfirth Estuary in 1974

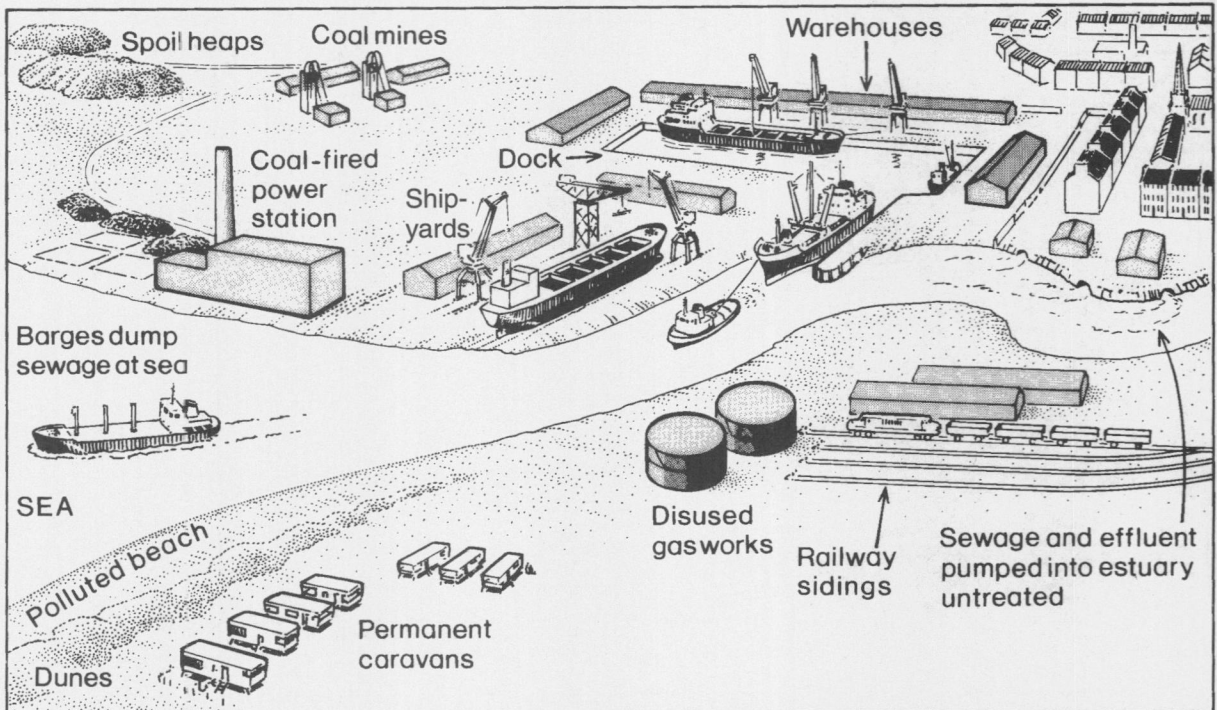

Spoil heaps Coal mines Warehouses

Coal-fired power station

Dock →

Ship-yards

Barges dump sewage at sea

SEA

Polluted beach

Dunes

Permanent caravans

Disused gas works

Railway sidings

Sewage and effluent pumped into estuary untreated

Reference Diagram Q6B: The Inverfirth Estuary in 2004

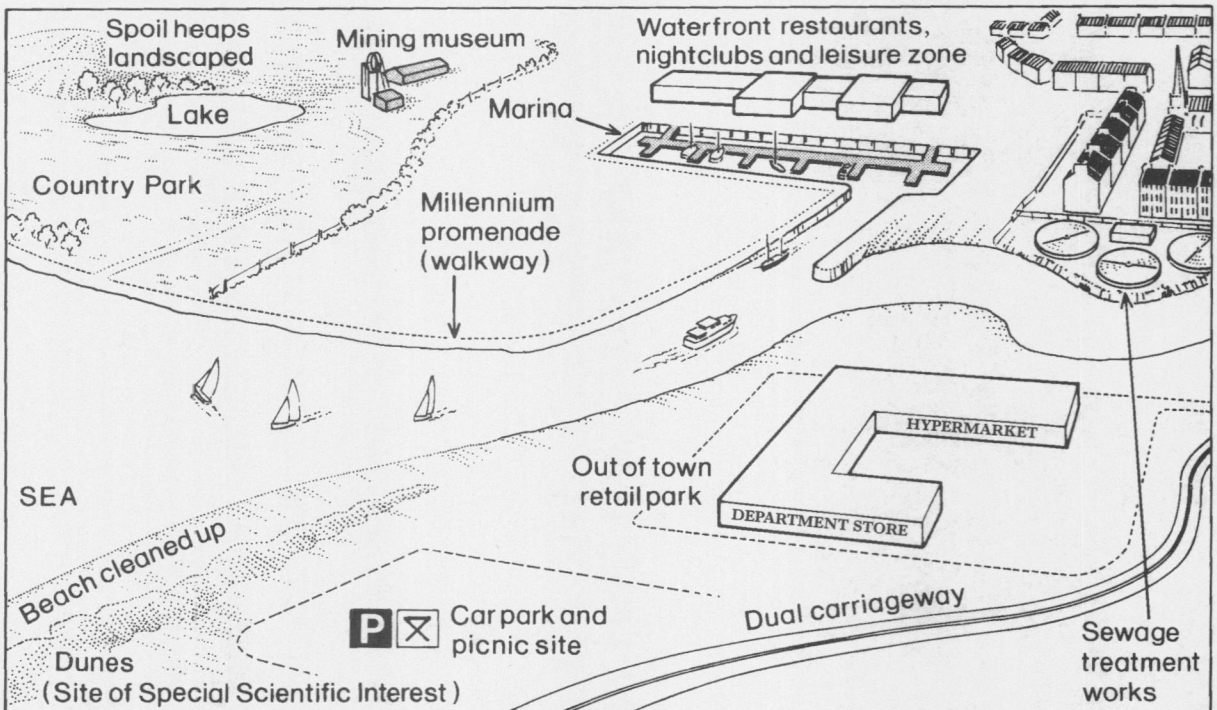

Spoil heaps landscaped

Mining museum

Waterfront restaurants, nightclubs and leisure zone

Lake

Marina

Country Park

Millennium promenade (walkway)

SEA

Beach cleaned up

HYPERMARKET

Out of town retail park

DEPARTMENT STORE

Dual carriageway

Sewage treatment works

P ⊠ Car park and picnic site

Dunes (Site of Special Scientific Interest)

DO NOT
WRITE IN
THIS
MARGIN

KU	ES

6. (continued)

Marks

(*a*)   Study Reference Diagrams Q6A and Q6B.

What techniques could pupils have used to gather the information shown on Reference Diagrams Q6A and Q6B?

Give reasons for your choice of techniques.

Techniques _____

_____

_____

Reasons _____

_____

_____

_____

_____

4

(*b*)   Do you think the changes that have taken place between 1974 and 2004 have improved the area?

Give reasons for your answer.

_____

_____

_____

_____

_____

_____

4

[Turn over

Marks

7.   **Reference Diagram Q7A: Shopping Centres in a large Town**

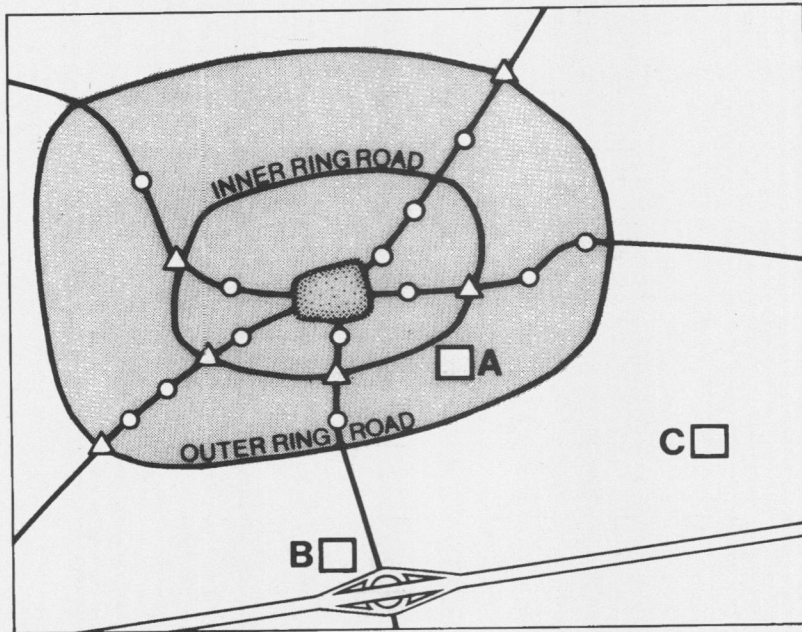

KEY

Built-up area

CBD

—— Main road

═══ Motorway

○ Small local centre

△ Large district centre

□ New retail park

(*a*)   Look at Reference Diagram Q7A above.

A, B and C are proposed sites for the development of a large retail park.

Which site do you think is best for this development? Give reasons for your choice.

Site _____

Reasons _____

_____

_____

_____

_____

4

DO NOT
WRITE IN
THIS
MARGIN

KU ES

Marks

7. (continued)

Reference Diagram Q7B: Employment Changes in the UK

		1980	1990	2000
Sectors of Industry (numbers in millions)	Primary	0·6	0·6	0·5
	Secondary	6·7	5·3	4·3
	Tertiary	16·8	20·1	22·3
	Total	24·1	26·0	27·1

(*b*) Give **two** processing techniques which could be used to present the information shown in Reference Diagram Q7B.

Give reasons for your choices.

Technique 1 _____

Reason _____

_____

_____

Technique 2 _____

Reason _____

_____

_____

4

[Turn over

Marks

8.   **Reference Diagram Q8: Population Pyramids**

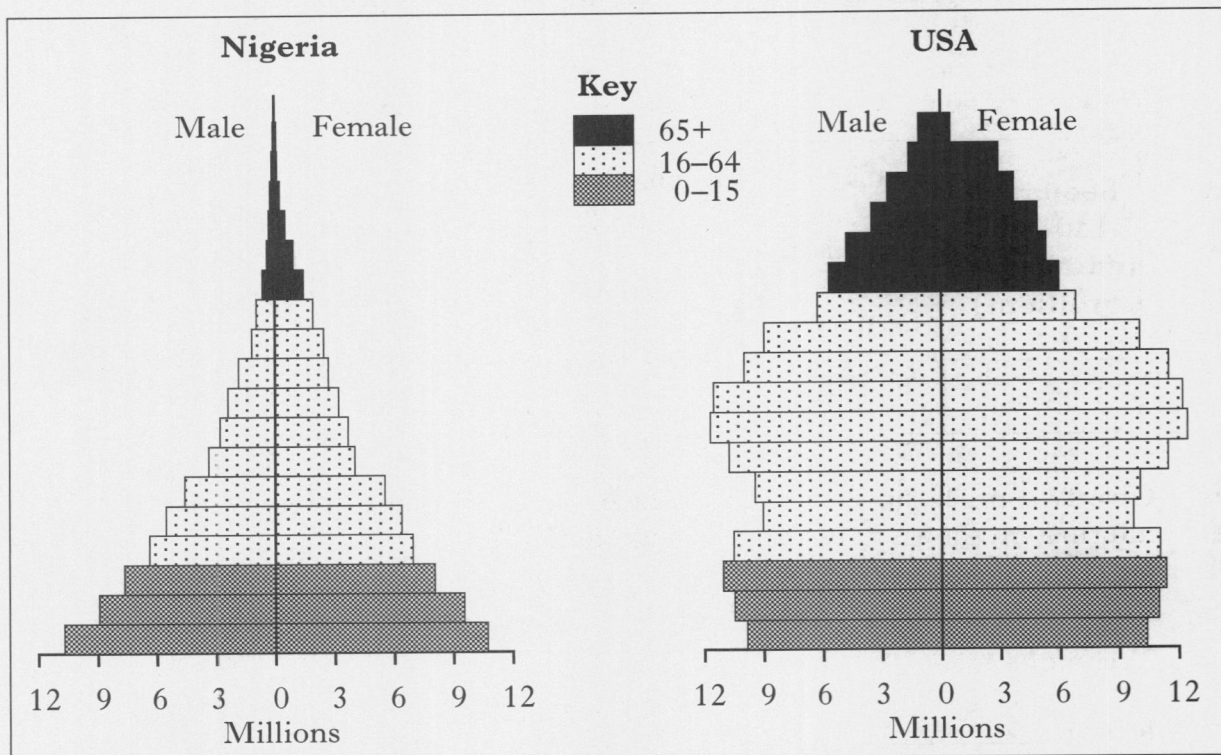

Nigeria

Male   Female

Key

65+
16–64
0–15

USA

Male   Female

12 9 6 3 0 3 6 9 12
Millions

12 9 6 3 0 3 6 9 12
Millions

(*a*)   Look at Reference Diagram Q8.

Describe three differences between the population pyramids of
Nigeria and the United States of America.

_____

_____

_____

_____   3

8. (continued)

Marks

(b) (i) **Describe** two problems which the population structure of Nigeria may cause.

_____

_____

_____

_____

(ii) **Describe** two problems which the population structure of the United States may cause.

_____

_____

_____

_____

4

[Turn over for Question 9 on *Page twenty*

KU	ES

Marks

9. **Reference Diagram Q9A: Factors Affecting Sugar Production in the EU**

Sugar beet

* high subsidies to farmers to produce beet sugar in the EU
* EU buys sugar in bulk from Mozambique
* import tariffs protect EU farmers from foreign competition
* good climate for growing sugar beet

Reference Diagram Q9B: Factors Affecting Sugar Production in Mozambique

Sugar cane

* lowest production costs in the world
* Mozambique has to sell sugar to EU at low prices
* cannot sell processed (refined) sugar to EU because of high import tariffs on processed food
* good growing conditions for sugar cane

Give reasons why sugar producers in Mozambique are at a **disadvantage** compared with sugar producers in the EU.

_____

_____

_____

_____

4

_____

[END OF QUESTION PAPER]

[BLANK PAGE]

C

1260/405

NATIONAL QUALIFICATIONS 2004

MONDAY, 17 MAY 1.00 PM – 3.00 PM

GEOGRAPHY STANDARD GRADE
Credit Level

All questions should be attempted.

Candidates should read the questions carefully. Answers should be clearly expressed and relevant.

Credit will always be given for appropriate sketch-maps and diagrams.

Write legibly and neatly, and leave a space of about one cm between the lines.

Marks may be deducted for bad spelling and bad punctuation, and for writing that is difficult to read.

All maps and diagrams in this paper have been printed in black only: no other colours have been used.

SCOTTISH QUALIFICATIONS AUTHORITY

©

Extract No 1349/EXP272

1:25 000 Scale
Explorer Series

Four colours should appear above; if not then please return to the invigilator.
Four colours should appear above; if not then please return to the invigilator.

Scale 1: 25 000

4 centimetres to 1 kilometre (one grid square)

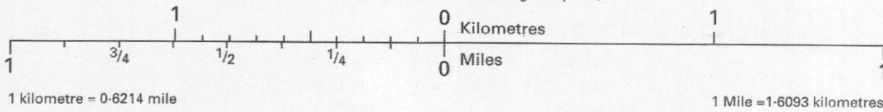

Kilometres

Miles

1 kilometre = 0·6214 mile

1 Mile =1·6093 kilometres

Magnetic North

Grid North

True North

Diagrammatic only

Reference Diagram Q1A

1.

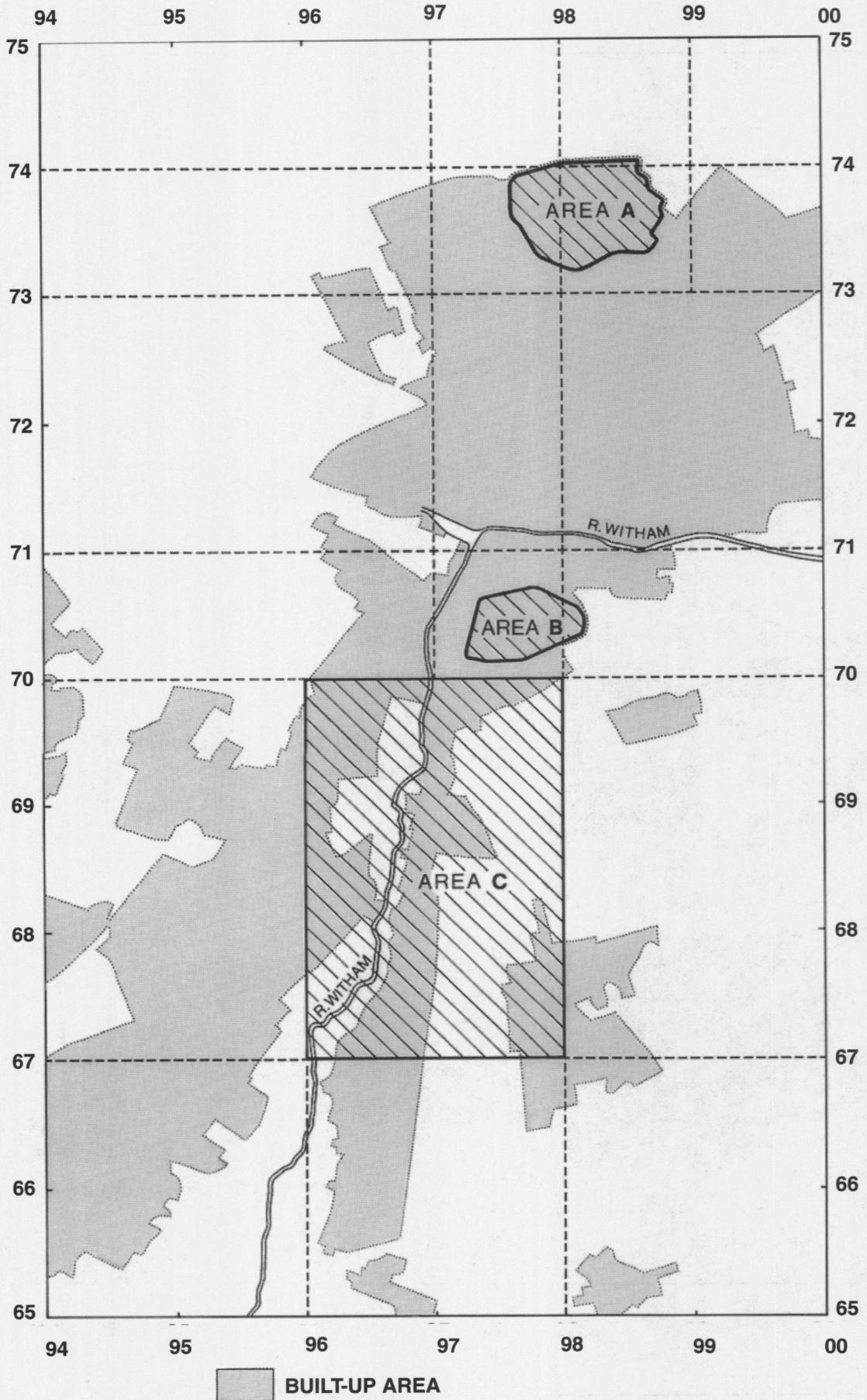

BUILT-UP AREA

Marks

KU	ES

1. (continued)

This question refers to the OS Map Extract (No 1349/EXP272) of Lincoln and the Reference Diagram Q1A on *Page two*.

(*a*) Give the Grid Reference of the square which contains the CBD of Lincoln. Support your answer with map evidence.

3

Look at Reference Diagram Q1A.

(*b*) Referring to map evidence, describe the differences between the residential environments of Area A and Area B.

4

(*c*)

> **"A dormitory settlement is a community where most of the residents travel to work in a larger settlement."**

Pupils from a local high school want to find out if Bracebridge Heath (9767–9867) is a dormitory settlement for Lincoln. What techniques could they use to gather relevant information?

Explain the choice of techniques.

5

(*d*) Give reasons for the differences between the leisure activities located in square 9771 and those located in squares 9468 and 9469.

4

(*e*) Suggest the type of farming found at Canwick Manor Farm (993677).

Give reasons for your choice.

4

(*f*) Referring to map evidence, explain how physical landscape features (relief and drainage) have affected land use in Area C.

6

[Turn over

Mar|
KU

2.   **Reference Diagram Q2A: Glaciated Upland**

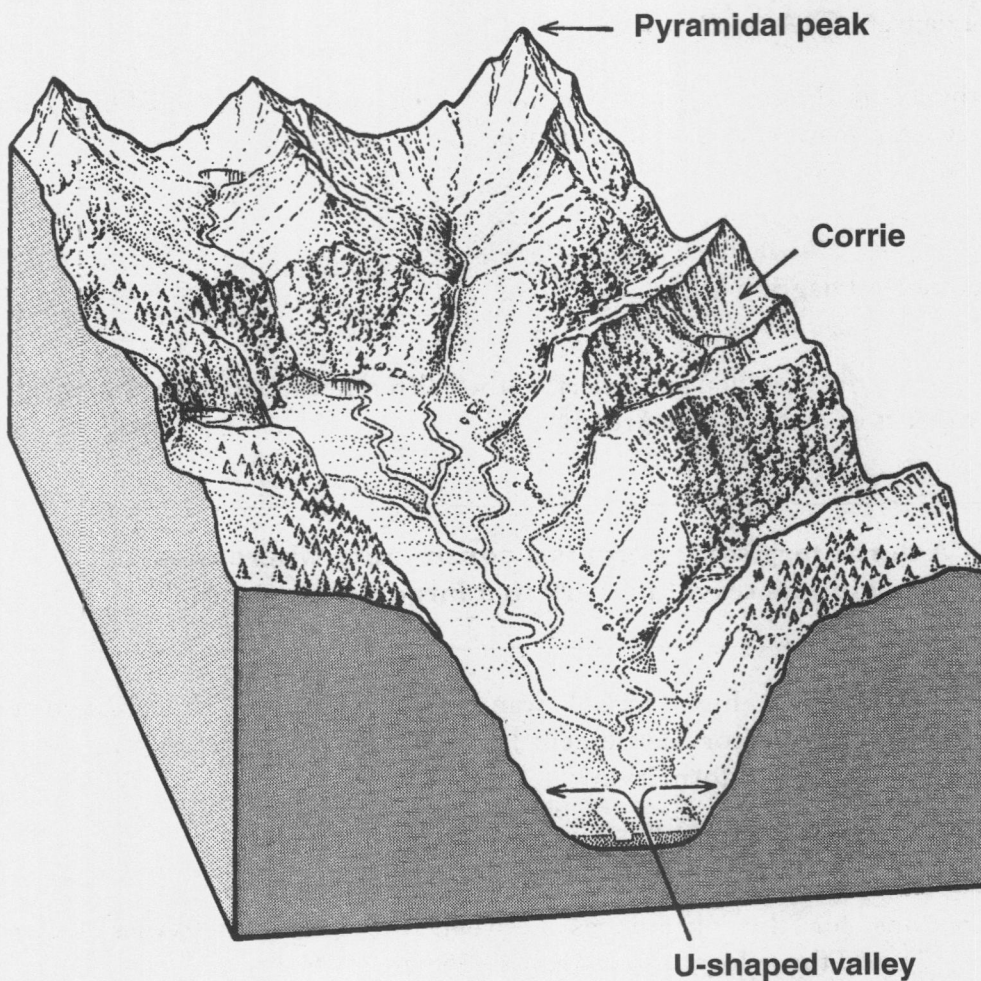

← **Pyramidal peak**

Corrie

U-shaped valley

(*a*) Study Reference Diagram Q2A.

Select **one** of the labelled glaciated features shown.

Explain in detail how it was formed. You may wish to use diagrams to illustrate
your answer.

4

Marks

KU	ES

2. (continued)

Reference Diagram Q2B: Some Land Uses in Glaciated Uplands

Forestry
Tourist Resorts
Skiing
Hydroelectric Power
Farming

(b) Select **one** land use shown in Reference Diagram Q2B.

Explain in detail why the land use is suited to a glaciated upland as shown in Reference Diagram Q2A.

4

[Turn over

3.

Reference Diagram Q3: Synoptic Chart for 25 September 2002

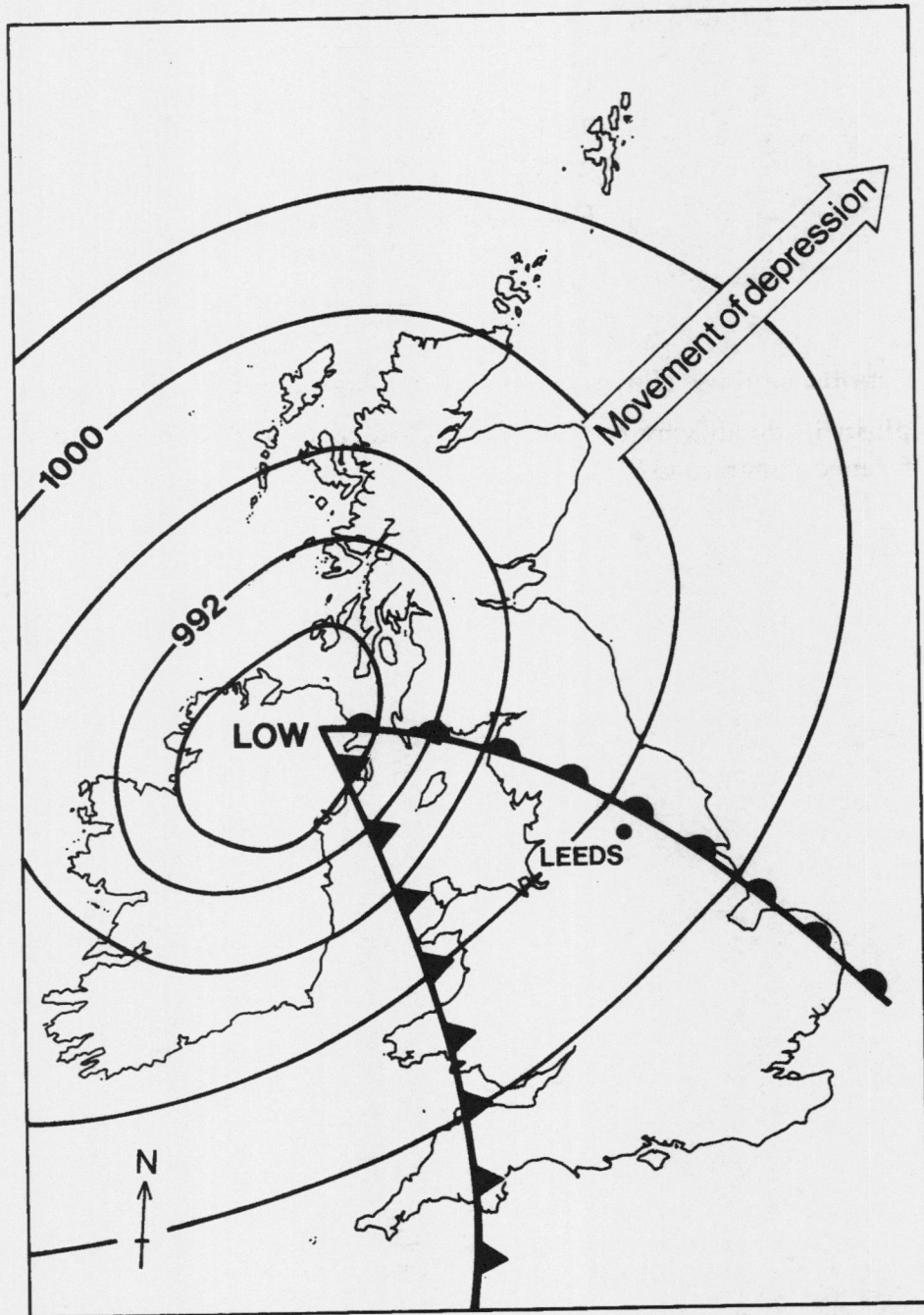

Study Reference Diagram Q3.

Explain the changes which will take place in the weather at Leeds in the next 12 hours.

5

4. **Reference Diagram Q4: Threats to the Marine Environment around Scotland**

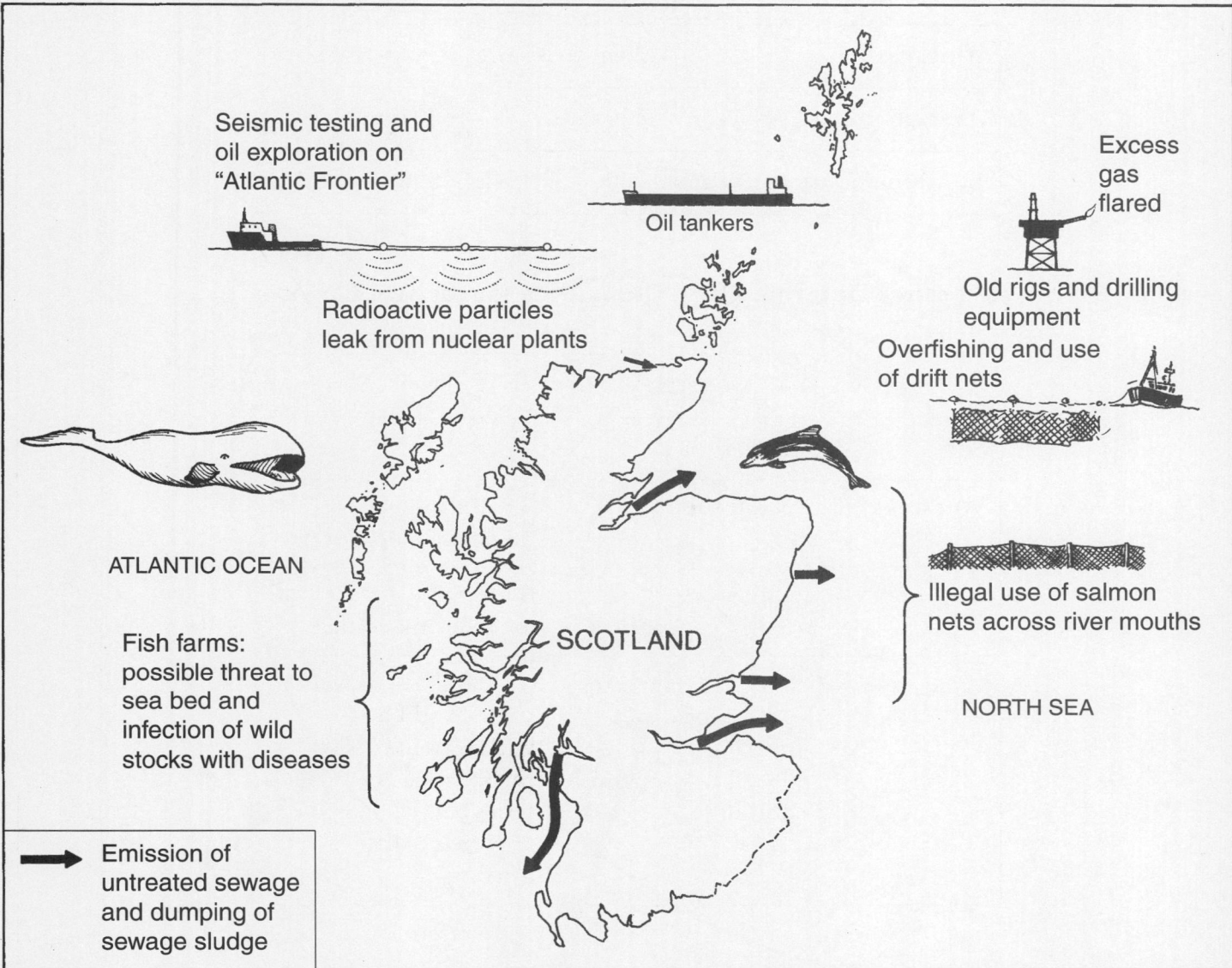

Seismic testing and oil exploration on "Atlantic Frontier"

Oil tankers

Excess gas flared

Radioactive particles leak from nuclear plants

Old rigs and drilling equipment

Overfishing and use of drift nets

ATLANTIC OCEAN

Fish farms: possible threat to sea bed and infection of wild stocks with diseases

SCOTLAND

Illegal use of salmon nets across river mouths

NORTH SEA

→ Emission of untreated sewage and dumping of sewage sludge

	Marks	
	KU	ES

What measures could be taken to reduce the impact of the threats to the marine environment as shown on Reference Diagram Q4?

6

[Turn over

Marks
KU

5. **Reference Diagram Q5A: Physical Data for two Farms in Scotland**

	Farm A	Farm B
Altitude	220 m to 450 m	75 m to 125 m
Average rainfall per year	1520 mm	630 mm
Sunshine hours per year	1000	1300

Reference Diagram Q5B: Other Data for the two Farms

	Farm A	Farm B
Area	1904 ha	444 ha
Workers	3 full time	5 full time 13 part time/seasonal
Machinery	2 tractors 8 other machines	6 tractors 14 other machines
Land use	80% sheep grazing 13% beef cattle grazing 7% barley, turnips and hay	87% arable—mainly wheat and barley with some potatoes, raspberries and strawberries 13% beef cattle grazing

(a) Look at Reference Diagrams Q5A and Q5B.

Give reasons for the differences between the two farms.

(b) Describe other techniques which could be used to present the land use data shown in Reference Diagram Q5B.

Give reasons for your choice of techniques.

6. Reference Diagram Q6A: Location of Toyota Car Factory at Burnaston

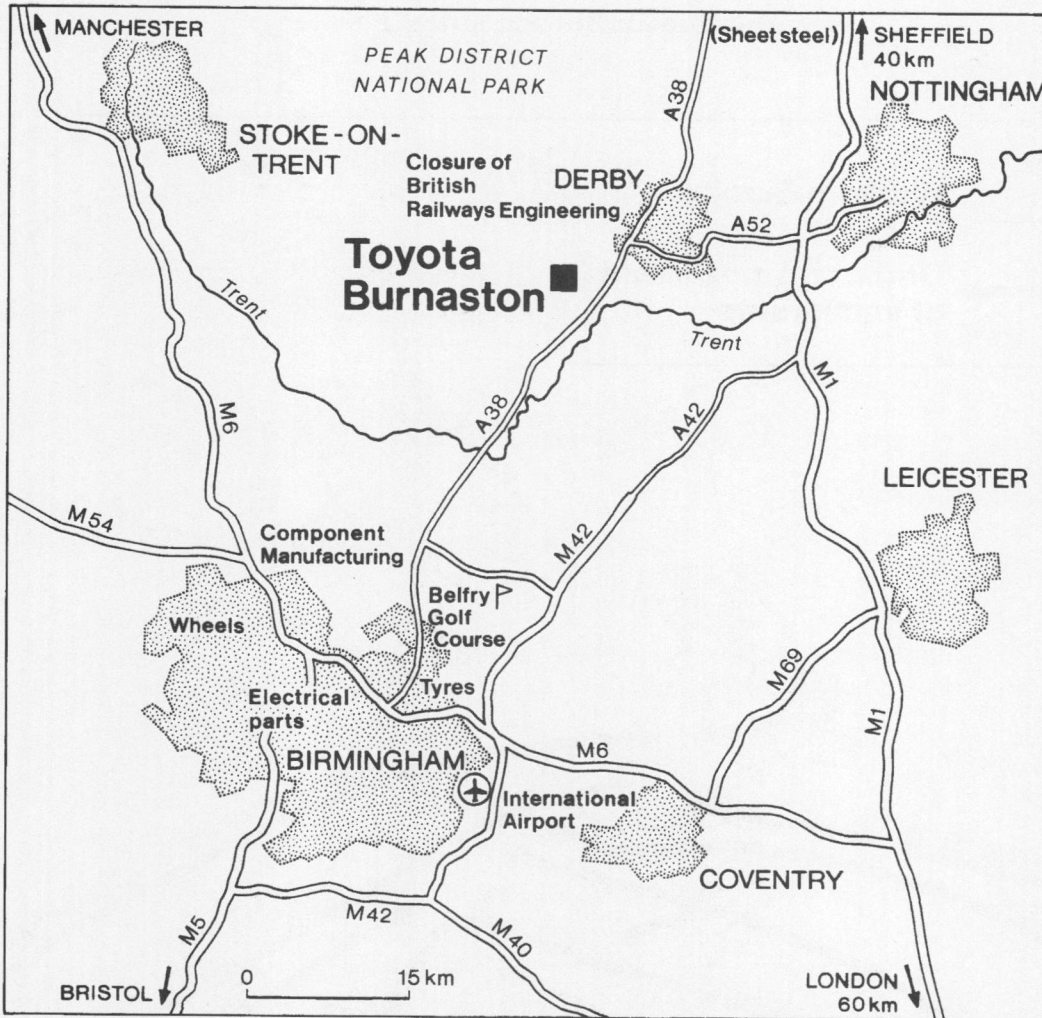

Reference Diagram Q6B: Site of Toyota's Burnaston Factory

Look at Reference Diagrams Q6A and Q6B.

What are the advantages of locating a car factory at Burnaston?

6

Marks

KU

7.   **Reference Diagram Q7A: Destination and Origins of Migrants into and within Europe since 1990**

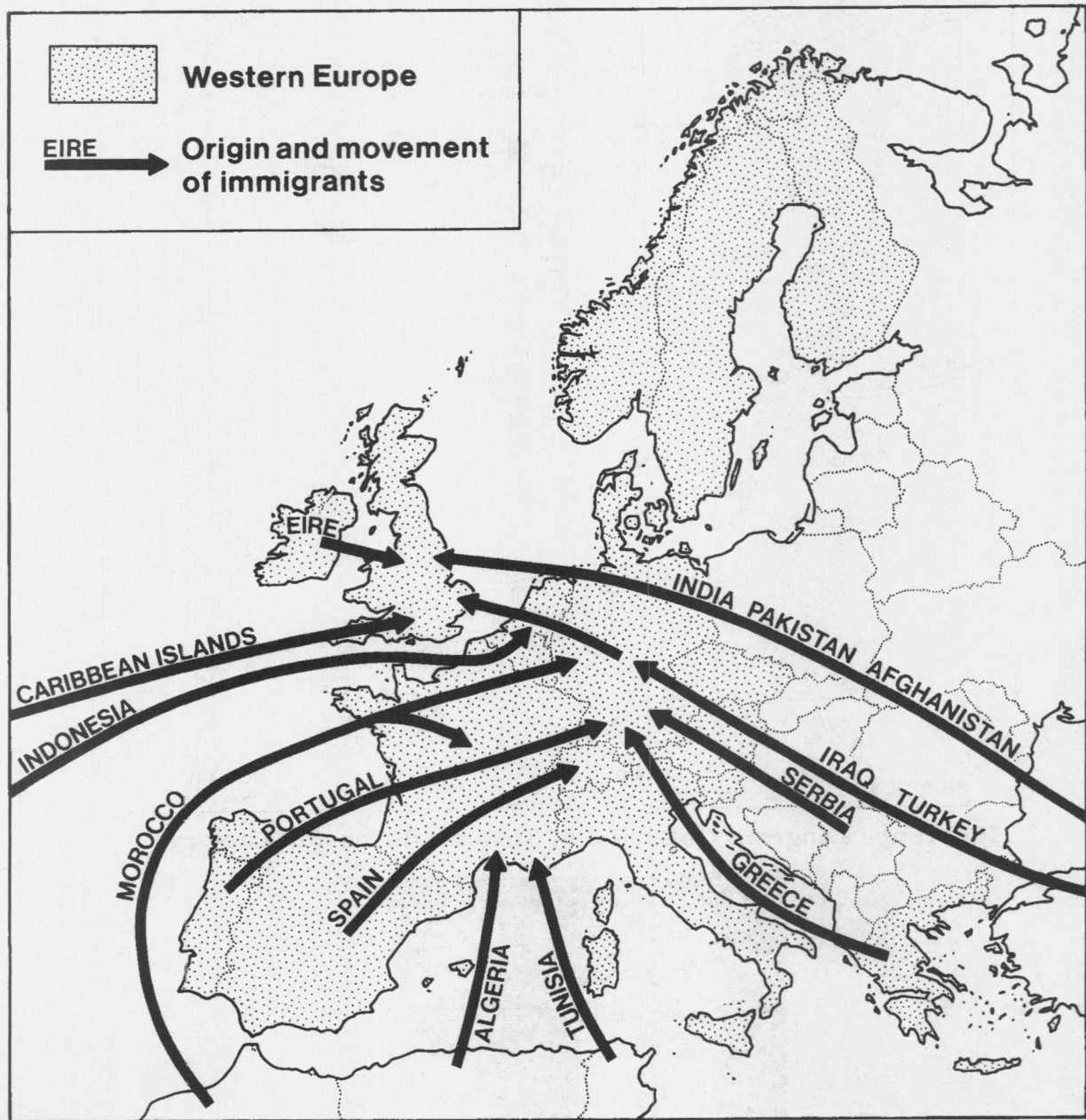

(*a*) Look at Reference Diagram Q7A above.

Describe the **pattern** of migration into and within Europe since 1990.

7. (continued)

Reference Diagram Q7B: Extract from Newspaper Article

Families from troubled countries given asylum

About 1200 Iraqi and Afghan people were given four-year work permits to live and work in the UK. One refugee said, "I can start work tomorrow. I have useful skills which can help me to provide for my family and put something back into this country."

Some local people say that they are very unhappy about the migrants moving to the UK.

"We thought we'd be safe in this country, but my family are still being persecuted," said one young mother from Kosovo; "Not everyone welcomes us."

(*b*) Look at Reference Diagram Q7B above.

What are the advantages **and** disadvantages **to migrants** of coming to countries in the European Union, such as the United Kingdom?

4

[Turn over

Mark

KU

8.   **Reference Diagram Q8: Location of the 10 new Members
of the European Union**

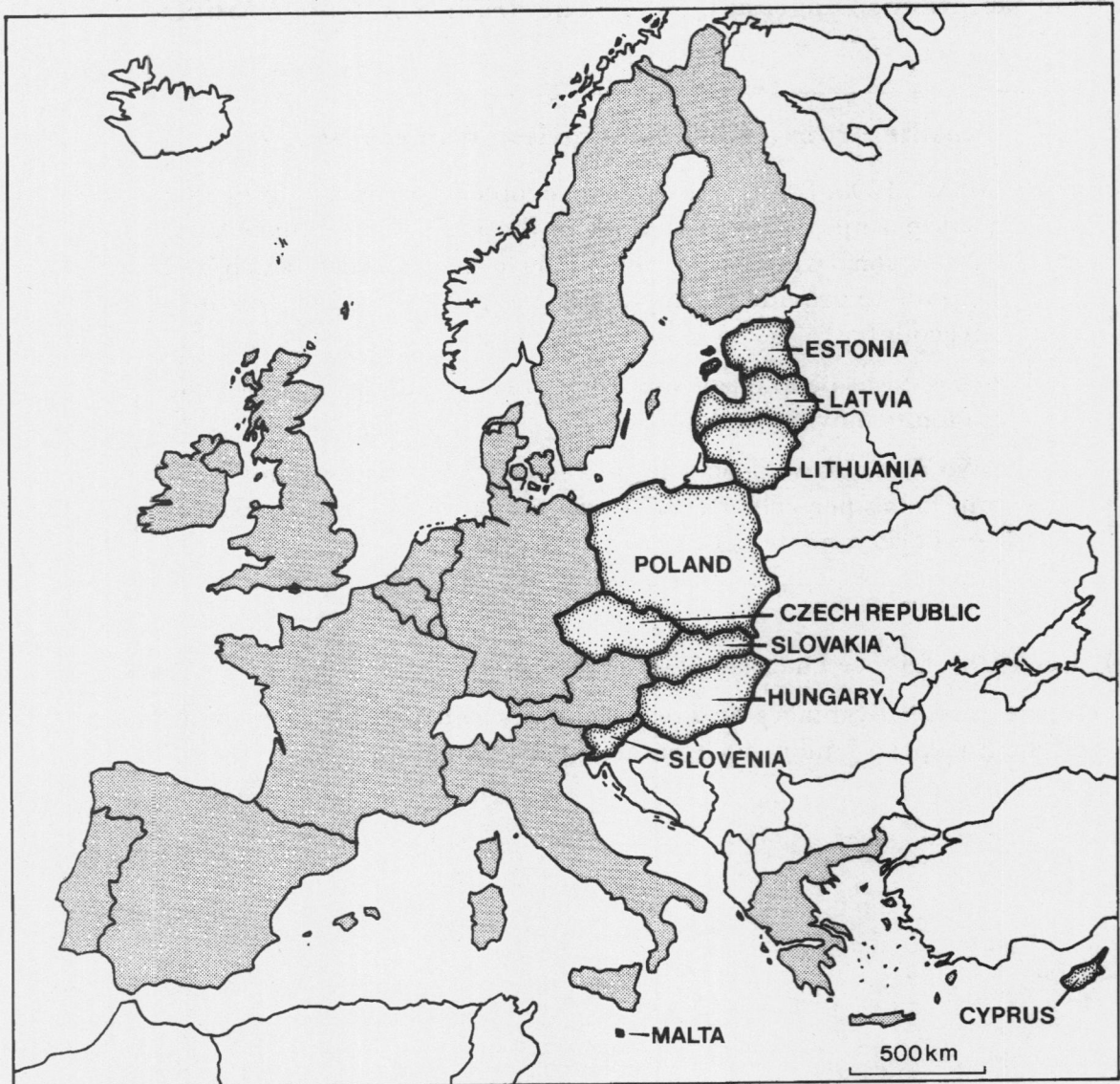

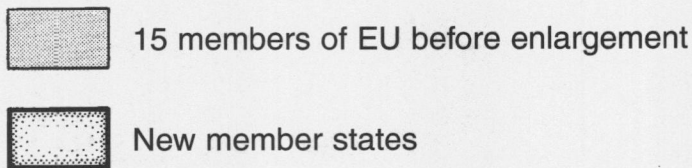

        15 members of EU before enlargement

        New member states

The European Union has been enlarged from its previous membership of 15 countries to a group of 25.

Explain the economic and political advantages to the 10 new countries of joining the European Union.

5

Marks

KU | ES

9.   **Reference Diagram Q9A: Mount Nyiragongo erupts, 17 January 2002**

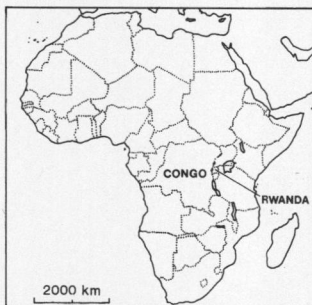

Reference Diagram Q9B

- dozens killed
- crops and farmland destroyed
- lava flows sweep through 14 villages, setting fire to fuel and power stations
- Goma Airport runway blocked by lava
- water supplies cut off
- 10 000 people made homeless
- harbour facilities on Lake Kivu destroyed

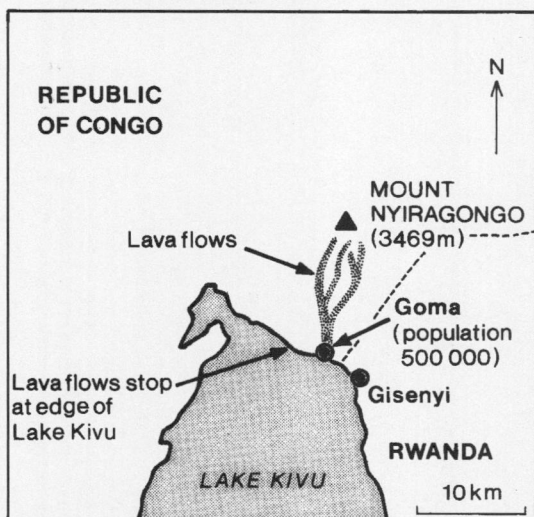

Reference Diagram Q9C

Short Term Aid	Long Term Aid
Tents and blankets	Road and bridge repairs
Medicines	New house building
Food supplies	Farming equipment and fertilisers

Look at Reference Diagrams Q9A, Q9B and Q9C.

Following the eruption of Mount Nyiragongo, aid was rushed to the area.

Which type of aid would be best suited to helping the people of this area following the volcanic eruption?

Give reasons for your answer.

5

[END OF QUESTION PAPER]

[BLANK PAGE]

[BLANK PAGE]

FOR OFFICIAL USE

G

KU | ES

Total Marks

1260/403

NATIONAL
QUALIFICATIONS
2005

WEDNESDAY, 11 MAY
10.25 AM–11.50 AM

GEOGRAPHY
STANDARD GRADE
General Level

Fill in these boxes and read what is printed below.

Full name of centre

Town

Forename(s)

Surname

Date of birth
Day Month Year

Scottish candidate number

Number of seat

1 Read the whole of each question carefully before you answer it.

2 Write in the spaces provided.

3 Where boxes like this ☐ are provided, put a tick ✓ in the box beside the answer you think is correct.

4 Try all the questions.

5 Do not give up the first time you get stuck: you may be able to answer later questions.

6 Extra paper may be obtained from the invigilator, if required.

7 Before leaving the examination room you must give this book to the invigilator. If you do not, you may lose all the marks for this paper.

SCOTTISH
QUALIFICATIONS
AUTHORITY

©

Ordnance Survey®

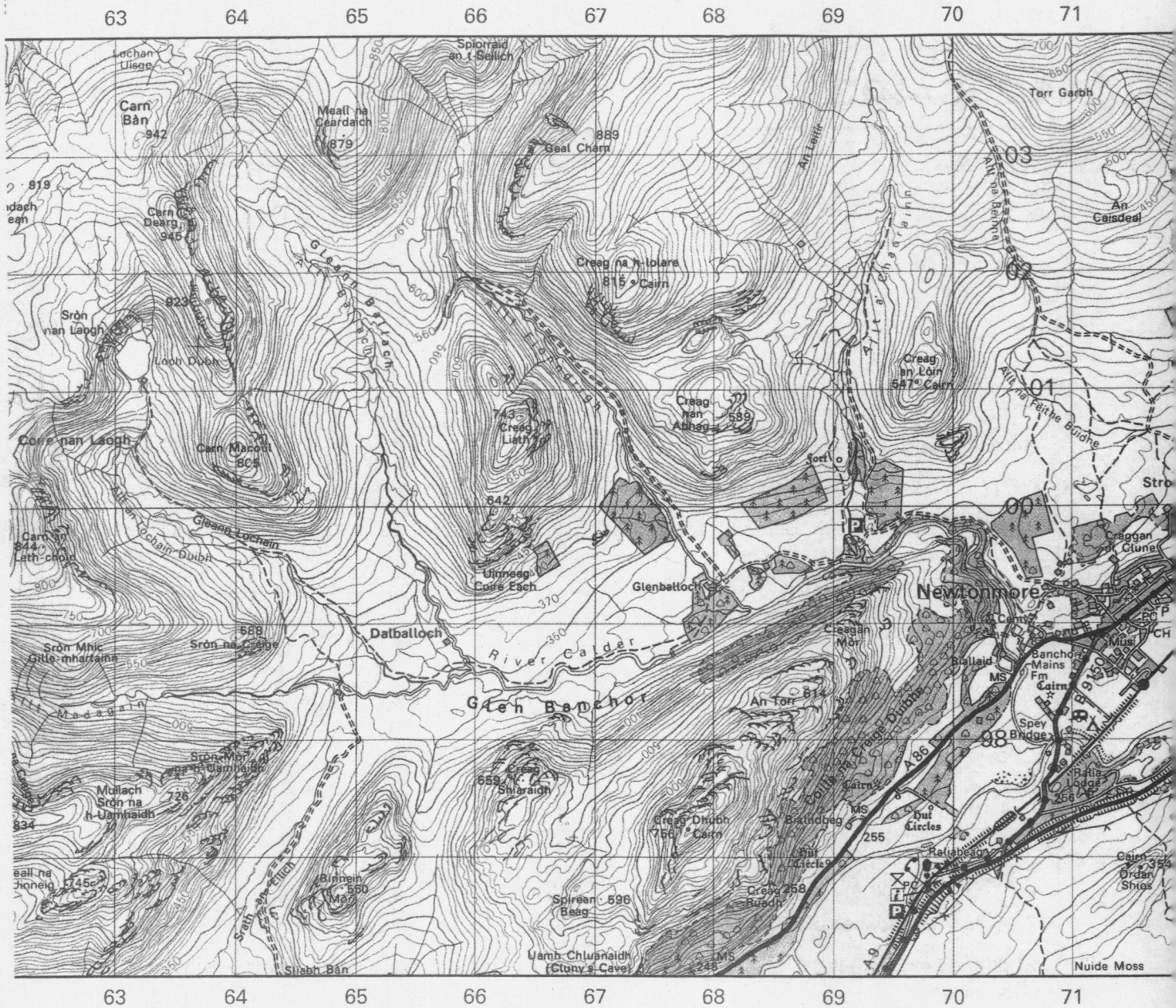

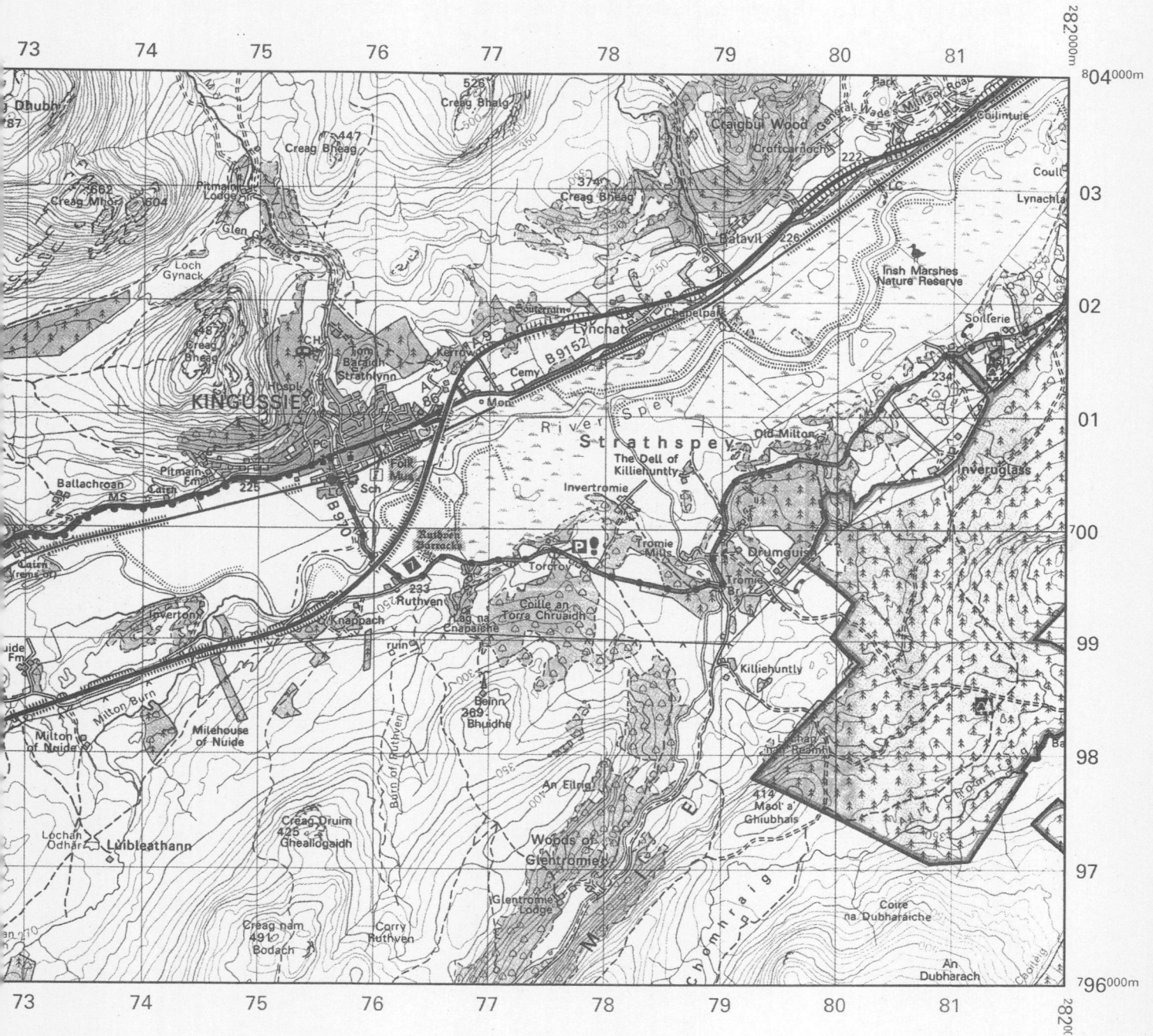

Extract No 1406/35

1:50 000 Scale
Landranger Series

Scale 1: 50 000

2 centimetres to 1 kilometre (one grid square)

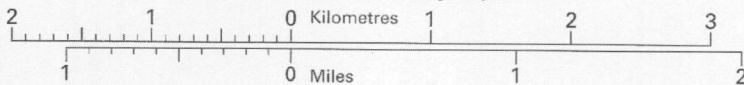

Kilometres

Miles

1 kilometre = 0·6214 mile

1 mile = 1·6093 kilometres

Magnetic North

Grid North

True North

Diagrammatic only

Marks

1. Look at the Ordnance Survey Map Extract (No 1406/35) of the Kingussie area.

(a) (i) Match the glacial features in the table with the grid references below. Choose from:

6399   6699   6301   7796.

Glacial Feature	Grid Reference
U-shaped valley with misfit stream	
Corrie with lochan	
Hanging valley	
Truncated spur with crags	

(ii) **Explain** how **one** of the glacial features named in the table above was formed. You may use a diagram(s) to illustrate your answer.

_____

_____

_____

_____

_____

3

3

KU	ES

Marks

1. (continued)

(b) The River Spey between 720981 and 820039 has a wide flood plain. Study this section of the River Spey. Using map evidence,

(i) give **two** ways in which people have made use of the flood plain;

_____

_____

_____

(ii) give **two** ways in which people have tried to overcome or avoid the problem of flooding on this section of the River Spey.

_____

_____

_____

4

(c) Pitmain Farm is in grid square 7400. What type of farm is this likely to be?

Tick (✓) your choice.

Livestock ☐   Mixed ☐   Arable ☐

Give reasons for your answer.

_____

_____

_____

_____

_____

3

[Turn over

DO NOT WRITE IN THIS MARGIN

KU ES

Marks

1. (continued)

(*d*) Match the site descriptions in the table with the settlements shown below:

Insh (8101) Newtonmore (7199) Kingussie (7500) Glenballoch (6799).

Description of Site	Name of Settlement
Between two tributaries to the north of the River Spey	
On land sloping gently down to the northwest, surrounded by forests	
On the floor of a U-shaped valley, next to a tributary of the River Calder	
A tributary of the River Spey runs through the middle of this settlement	

3

(*e*) Do you think the area around Newtonmore is suitable for tourism? Using map evidence, give reasons for your answer.

____

____

____

____

____

____

4

Marks

DO NOT WRITE IN THIS MARGIN

KU | ES

1. (continued)

(f)   **Reference Diagram Q1A: Sketch Map showing part of Insh Marshes Nature Reserve**

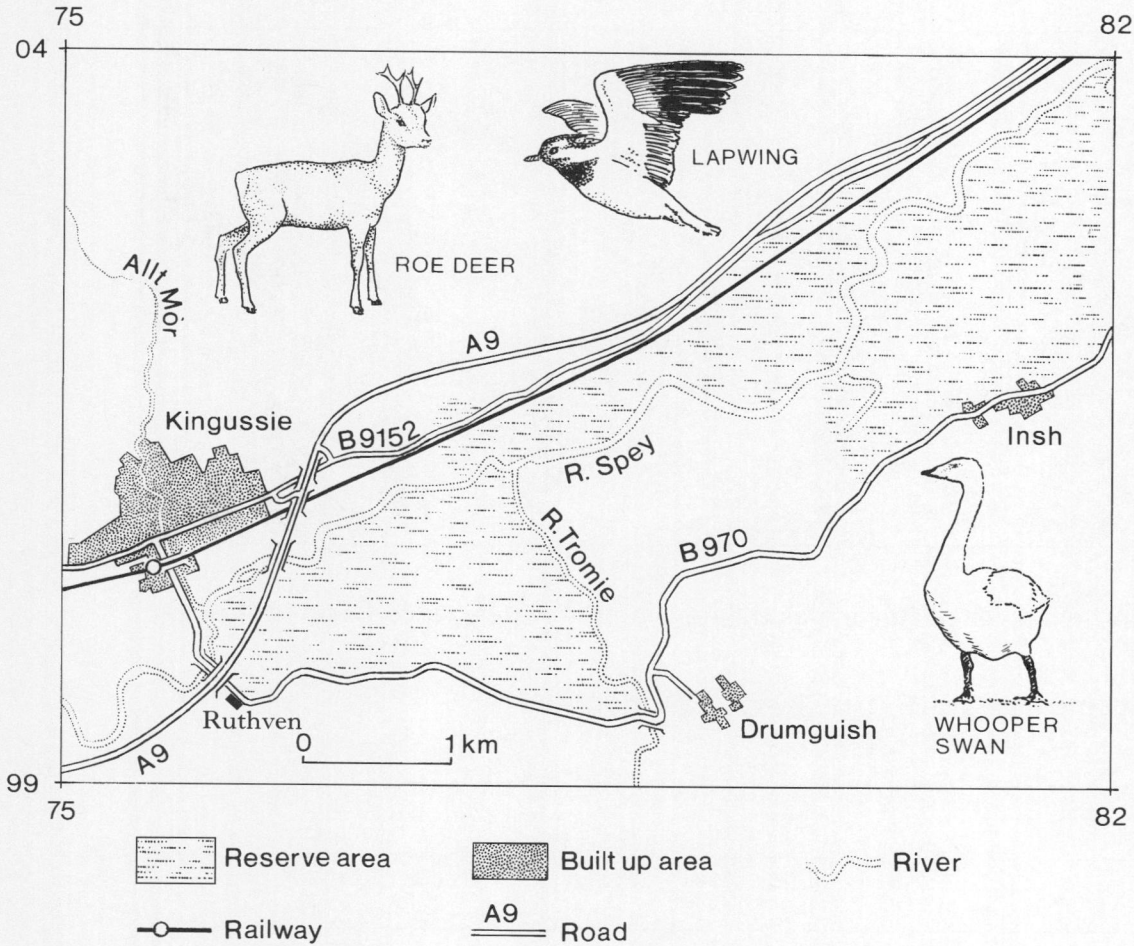

Legend:

- Reserve area
- Built up area
- ~~~ River
- —o— Railway
- A9 Road

Look at the Ordnance Survey map and Reference Diagram Q1A.

The Insh Marshes Nature Reserve extends on both sides of the River Spey from grid square 7699 (Ruthven) to grid square 8103.

Using map evidence, give advantages and disadvantages of the site of this nature reserve.

Advantages _____

_____

_____

_____

Disadvantages _____

_____

_____

4

KU ES

Marks

2.   **Reference Diagram Q2: Waterfall**

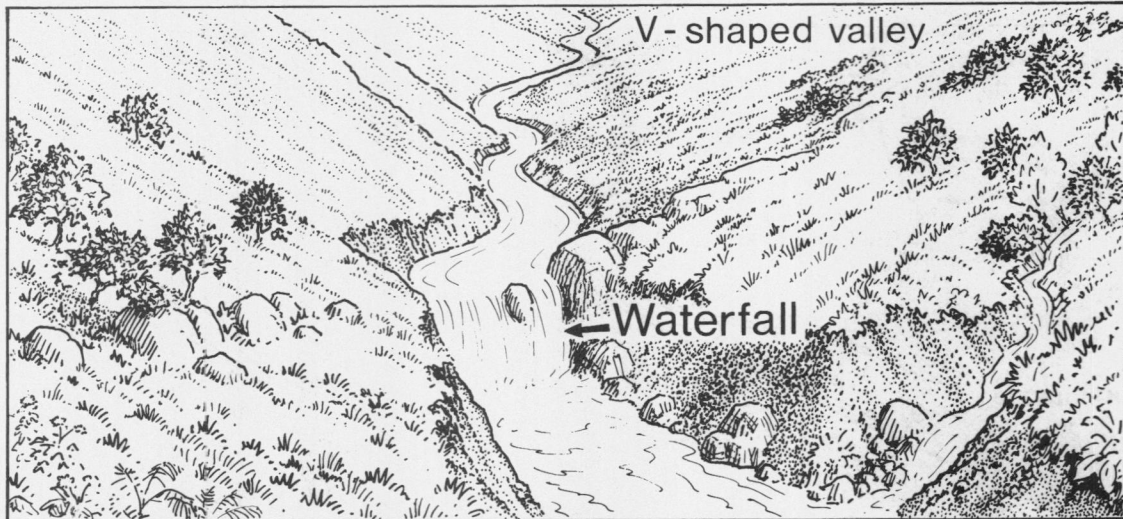

Explain how a waterfall such as the one shown in Reference Diagram Q2 is formed.

You may use a diagram(s) to illustrate your answer.

_____

_____

_____

_____

_____

KU ES

3

[Turn over for Question 3 on *Page eight*

Official SQA Past Papers: General Geography 2005

DO NOT
WRITE IN
THIS
MARGIN

KU ES

Marks

3. **Reference Diagram Q3A: Weather Station Sites**

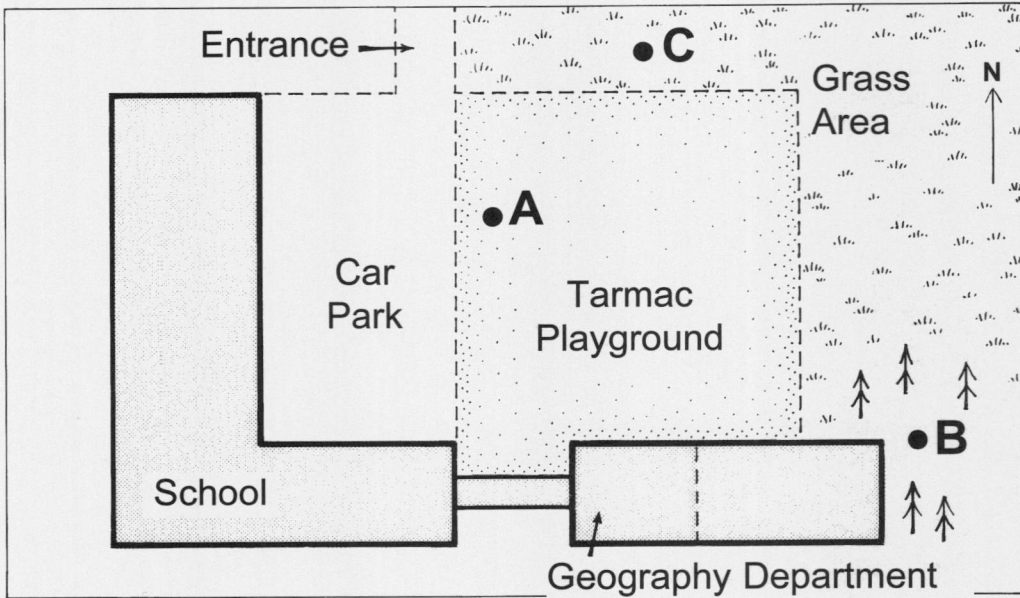

(a) Look at Reference Diagram Q3A.

Do you agree that site A is the best site for the school weather station?

Tick (✓) your choice.

Yes ☐   No ☐

Give reasons for your choice.

_____

_____

_____

_____

_____

3

Marks

3. **(continued)**

Reference Diagram Q3B: Air Masses affecting Britain

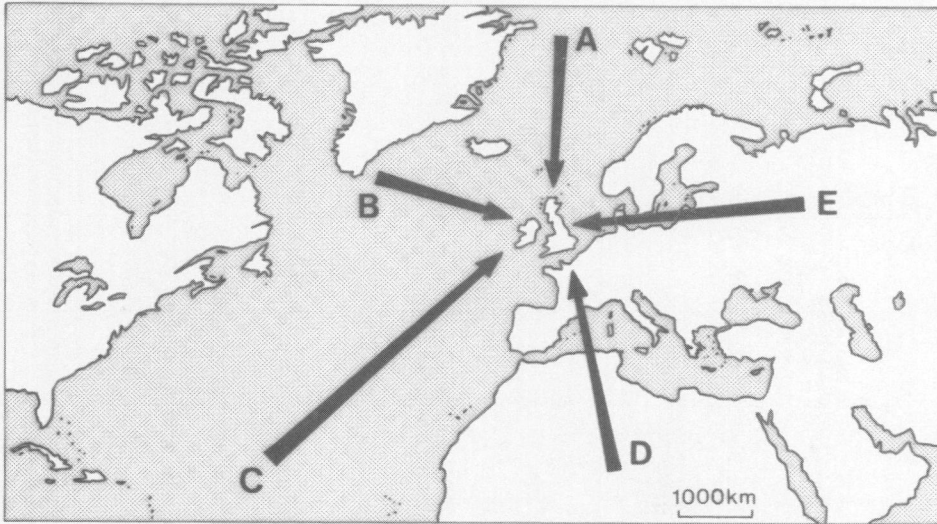

(b) Look at Reference Diagram Q3B.

(i) Describe the type of weather Britain might have with air mass D.

_____

_____

_____

_____

2

(ii) **Explain** why air mass B would bring cold, wet weather.

_____

_____

_____

_____

2

[Turn over

KU | ES

Marks

4.   **Reference Diagram Q4A: Selected Climate Regions**

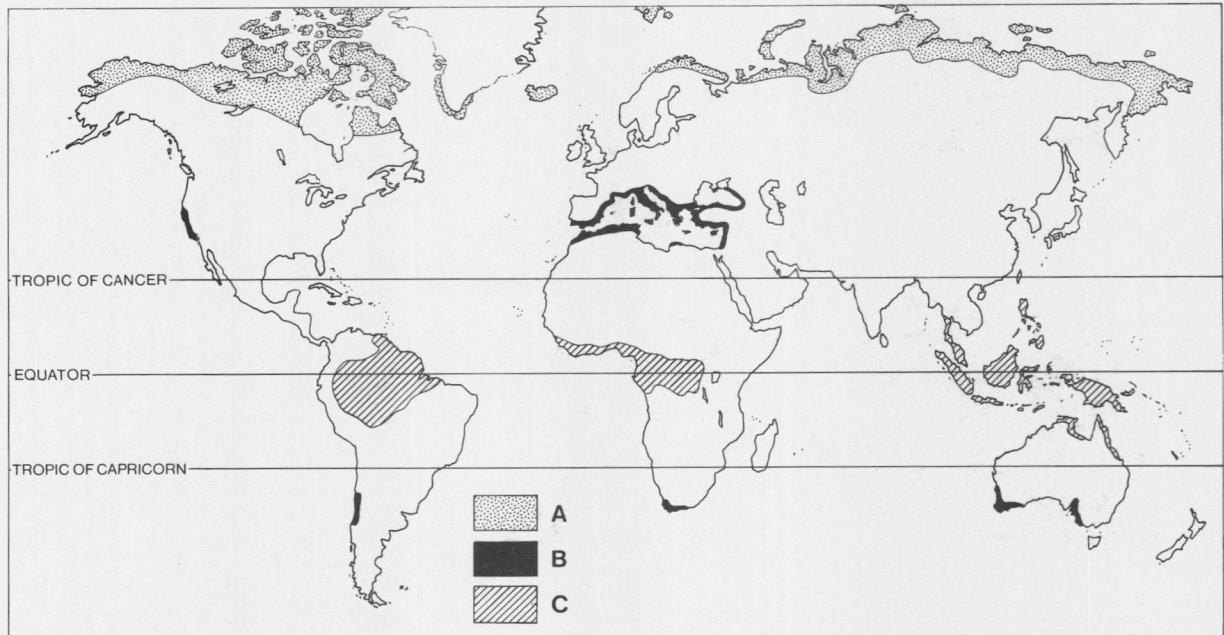

TROPIC OF CANCER

EQUATOR

TROPIC OF CAPRICORN

A

B

C

(a)   Look at Reference Diagram Q4A above.

Complete the table below by writing in the names of the climate regions.

Area	Name of Climate Region
A	
B	
C	

3

DO NOT
WRITE IN
THIS
MARGIN

KU | ES

Marks

4. **(continued)**

Reference Diagram Q4B: Desertification in Africa

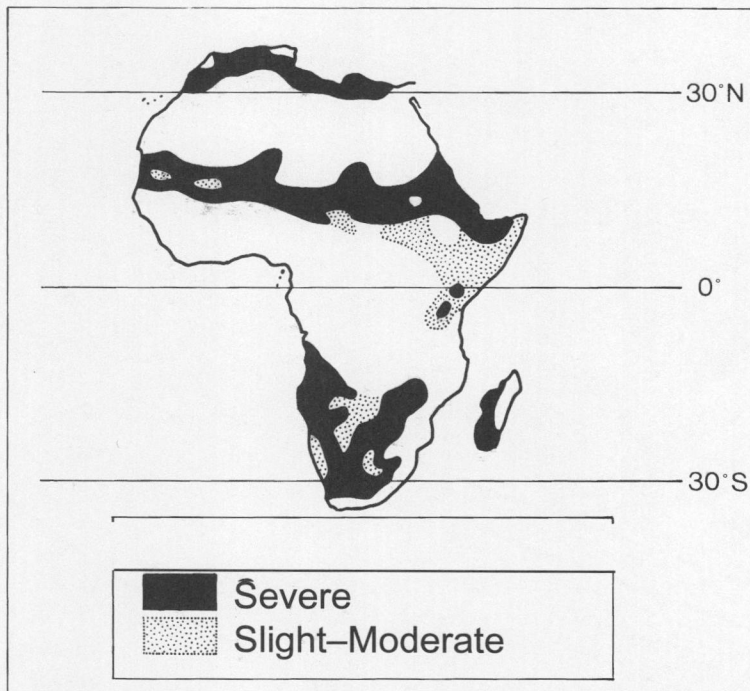

30°N

0°

30°S

■ Severe
⋮ Slight–Moderate

(*b*)   Look at Reference Diagram Q4B.

What are the main causes of desertification?

_____

_____

_____

_____

_____

_____

4

[Turn over

5.   **Reference Diagram Q5: Land Uses at the Edge of a City**

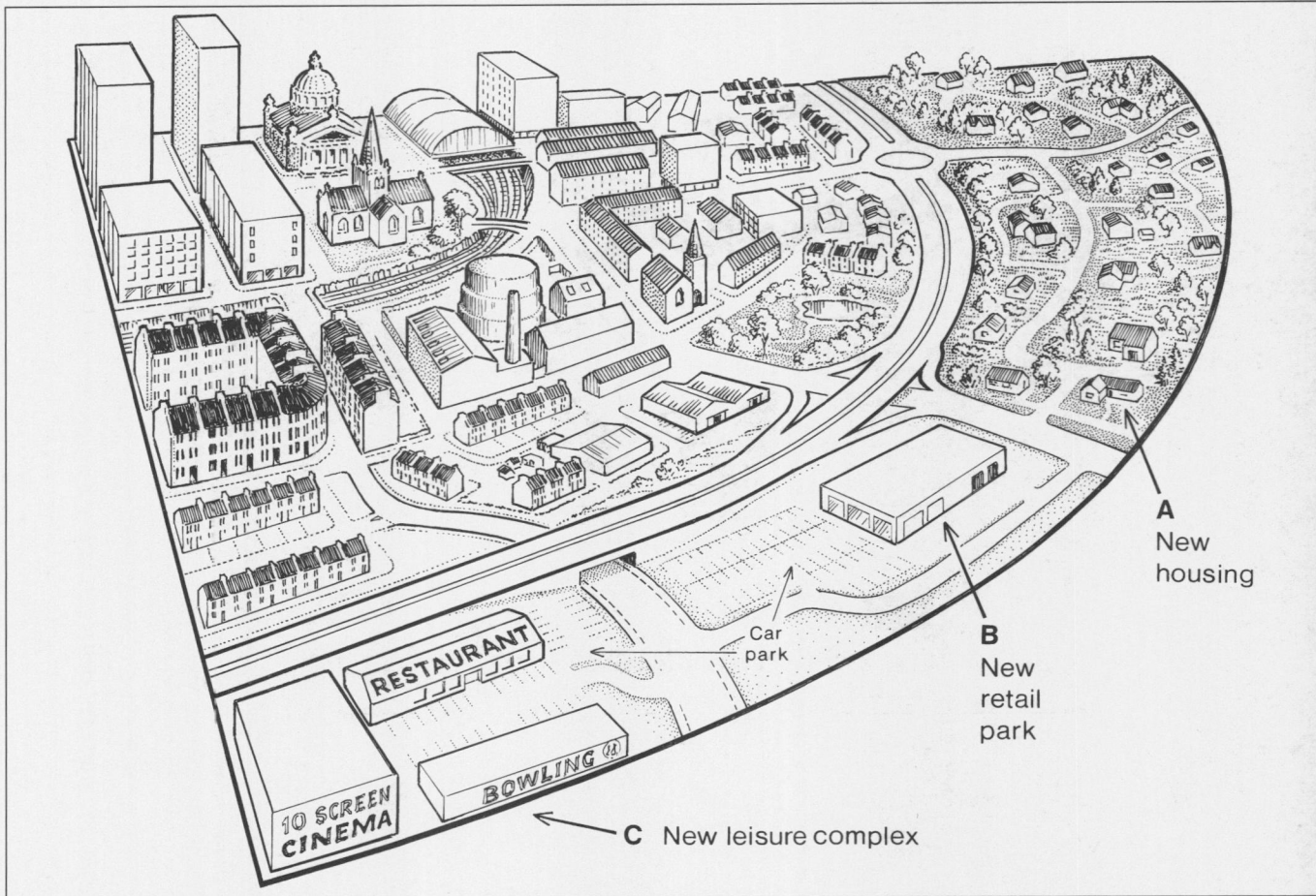

Look at Reference Diagram Q5.

(a)   Choose **one** of the main land uses in the diagram—A, B or C.

Chosen land use _____

Explain why the edge of the city is a good location for this land use.

_____

_____

_____

_____

_____

DO NOT WRITE IN THIS MARGIN

KU | ES

Marks

5. (continued)

(*b*) Identify **two** techniques which pupils could use to gather information in a study of an out-of-town shopping centre.

Give reasons for your choice.

Technique 1 ______________________________

Technique 2 ______________________________

Reasons for choice ______________________________

____

____

____

____

______________________________ **4**

[Turn over

Marks

6.　　　**Reference Diagram Q6: Recent Trends in Farming**

Organic
crops

Chemical
fertilisers

GM
crops

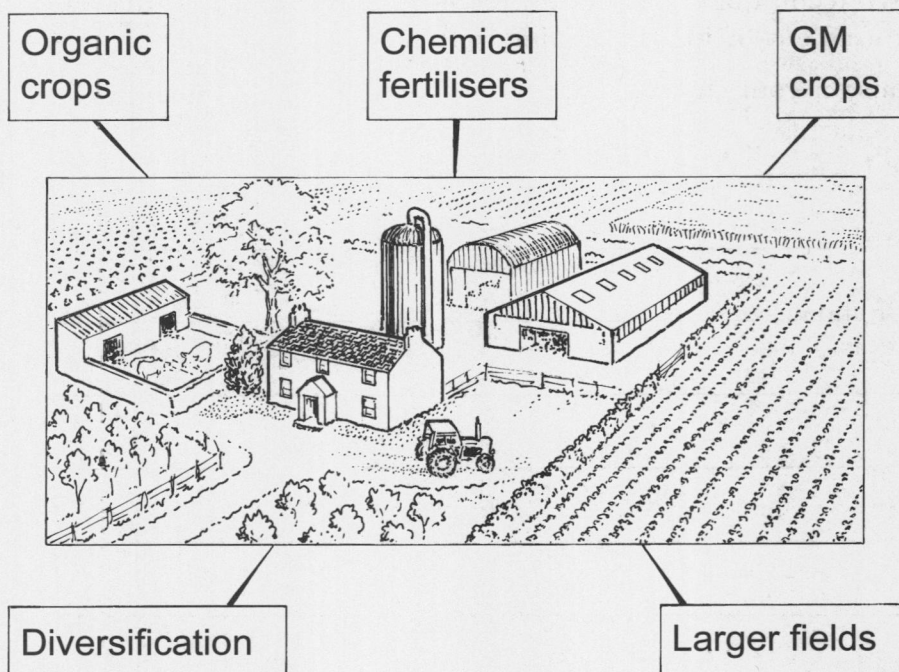

Diversification

Larger fields

"Recent trends in farming are of great benefit to the British people."

Do you agree? Explain your answer.

_____

_____

_____

_____

_____

_____

4

7. **Reference Diagram Q7: Old Industrial Area**

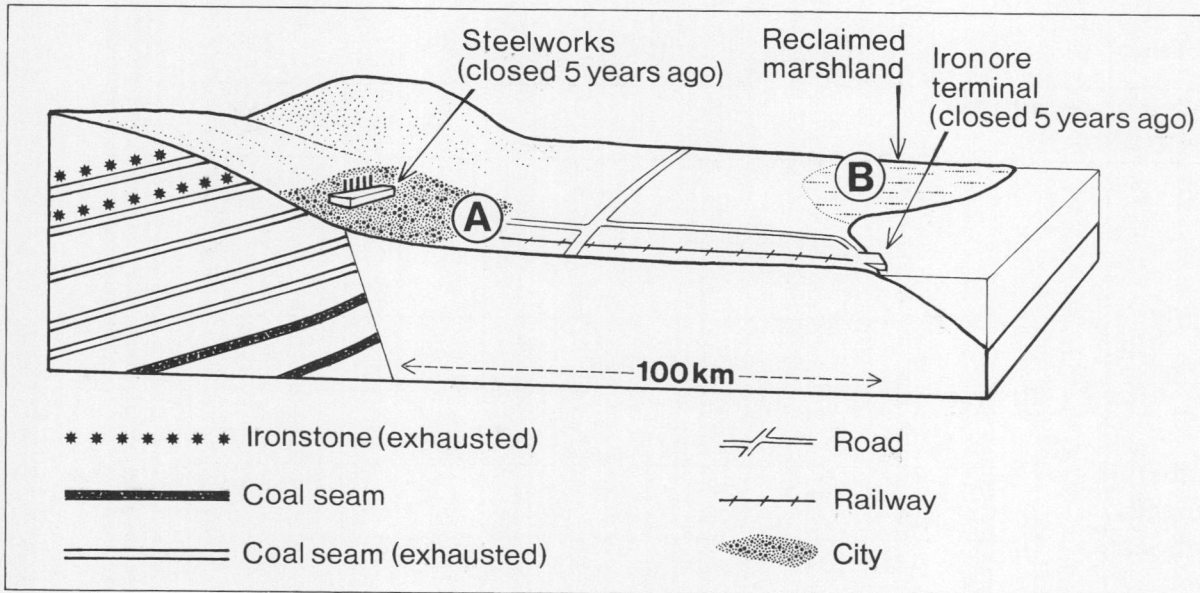

Steelworks (closed 5 years ago)

Reclaimed marshland

Iron ore terminal (closed 5 years ago)

Ⓐ

Ⓑ

←--------- **100 km** --------→

* * * * * * * Ironstone (exhausted)

▬▬▬▬▬ Coal seam

═══════ Coal seam (exhausted)

━╱━╱━ Road

━⊦━⊦━ Railway

▦ City

KU | ES

Marks

Look at Reference Diagram Q7.

A steel company is considering building a new integrated steelworks at either A or B.

Which site do you think they should choose?

Give reasons for your answer.

Chosen site _____

_____

_____

_____

_____

_____

_____

4

[Turn over

Marks

8.   **Reference Diagram Q8A: World Refugees (millions) 1971–2001**

Year	1971	1981	1991	2001
No of refugees (millions)	3	8	18	24

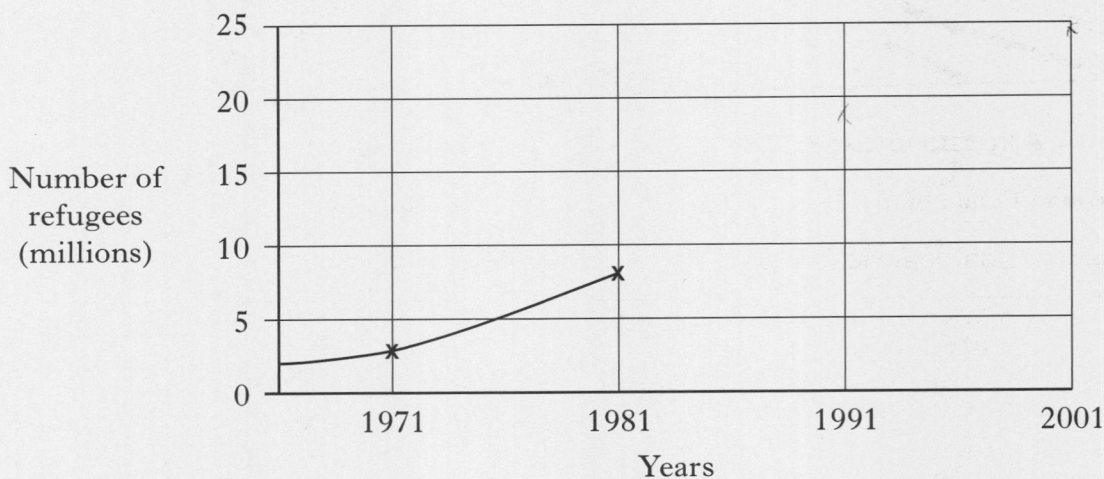

Number of refugees (millions)

25

20

15

10

5

0

1971   1981   1991   2001

Years

(*a*)   Complete the line graph using the figures from Reference Diagram Q8A.   **2**

KU | ES

Marks

8. (continued)

Reference Table Q8B: Sources of Refugees in Glasgow 2003

Country	Percentage
Afghanistan	6
Democratic Republic of Congo	4
Iran	9
Iraq	7
Pakistan	9
Somalia	9
Turkey	14
Others	42

(b) Suggest **two** other techniques which could be used to process the information in Reference Table Q8B.

Give reasons for your answers.

Technique 1 _____

Technique 2 _____

Reasons _____

_____

_____

_____

_____

_____

4

[Turn over

Marks

9.   **Reference Diagram Q9A: Quotation from Kofi Annan**

"Farmers in poor countries must be given a fair chance to compete, both in world markets and at home."

Kofi Annan (United Nations General Secretary)

Reference Diagram Q9B: Maize Trade between USA and Mexico

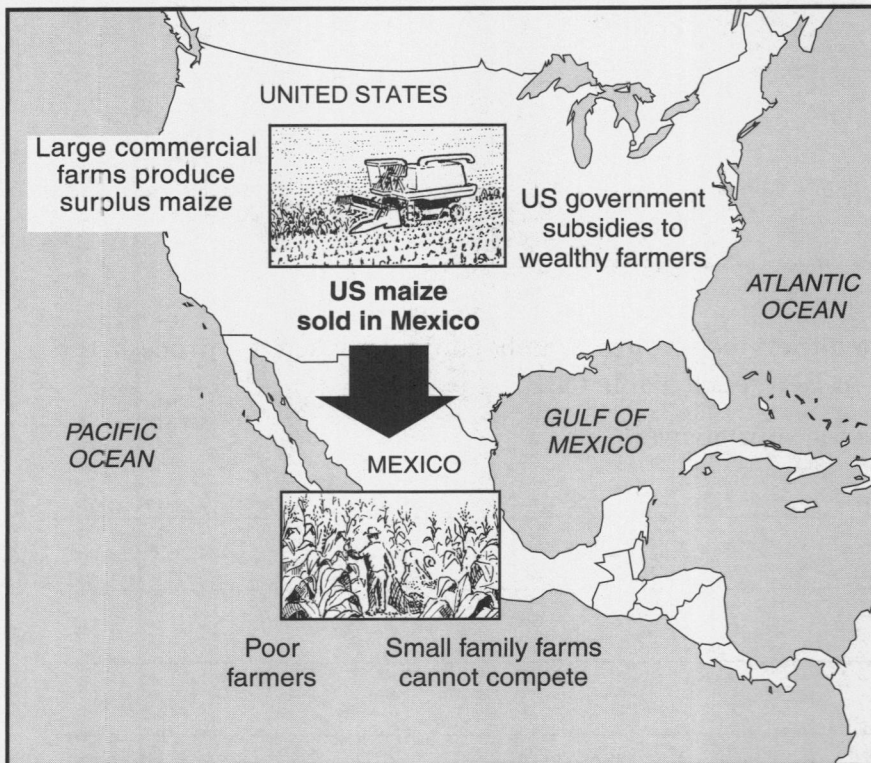

Look at Reference Diagrams Q9A and Q9B.

Explain why Mexican farmers think the maize trade between the USA and their country is unfair.

_____

_____

_____

_____

_____

_____

_____

4

[Turn over for Question 10 on *Page twenty*

Marks

10.

Reference Diagram Q10A: Mozambique

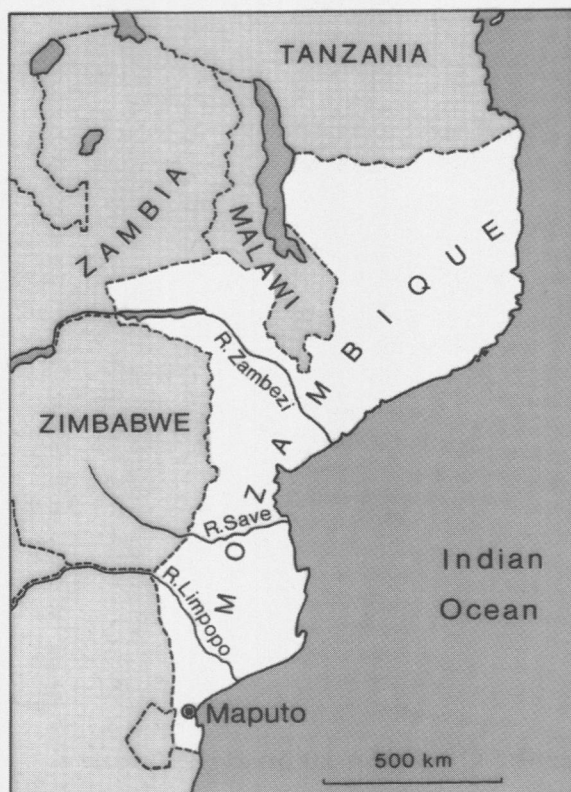

Reference Diagram Q10B: Effects of the Mozambique Floods, February/March 2000

Deaths	1000
Homeless	500 000
Farmland lost	25%
Schools destroyed	600
Estimated cost of rebuilding	$270–$430 million

Reference Diagram Q10C: Types of Aid

Short-term Aid	Long-term Aid
Clean water	Rebuilding homes
Food	Road building
Emergency shelter	Electricity network
Medicines	Building hospitals

DO NOT
WRITE IN
THIS
MARGIN

KU ES

Marks

10. (continued)

Look at Reference Diagrams Q10A, Q10B and Q10C.

In 2000, two cyclones hit Mozambique causing the Rivers Save, Zambezi and Limpopo to burst their banks. Almost half of Mozambique's land was flooded.

Which type of aid, short-term or long-term, would have been most useful to Mozambique?

Explain your answer in detail.

_____

_____

_____

_____

_____

_____

4

[END OF QUESTION PAPER]

[BLANK PAGE]

[BLANK PAGE]

C

1260/405

NATIONAL QUALIFICATIONS 2005	WEDNESDAY, 11 MAY 1.00 PM – 3.00 PM	**GEOGRAPHY STANDARD GRADE** Credit Level

All questions should be attempted.

Candidates should read the questions carefully. Answers should be clearly expressed and relevant.

Credit will always be given for appropriate sketch-maps and diagrams.

Write legibly and neatly, and leave a space of about one cm between the lines.

Marks may be deducted for bad spelling and bad punctuation, and for writing that is difficult to read.

All maps and diagrams in this paper have been printed in black only: no other colours have been used.

SCOTTISH QUALIFICATIONS AUTHORITY

Extract No 1407/EXP349

1:25 000 Scale
Explorer Series

Four colours should appear above; if not then please return to the invigilator.
Four colours should appear above; if not then please return to the invigilator.

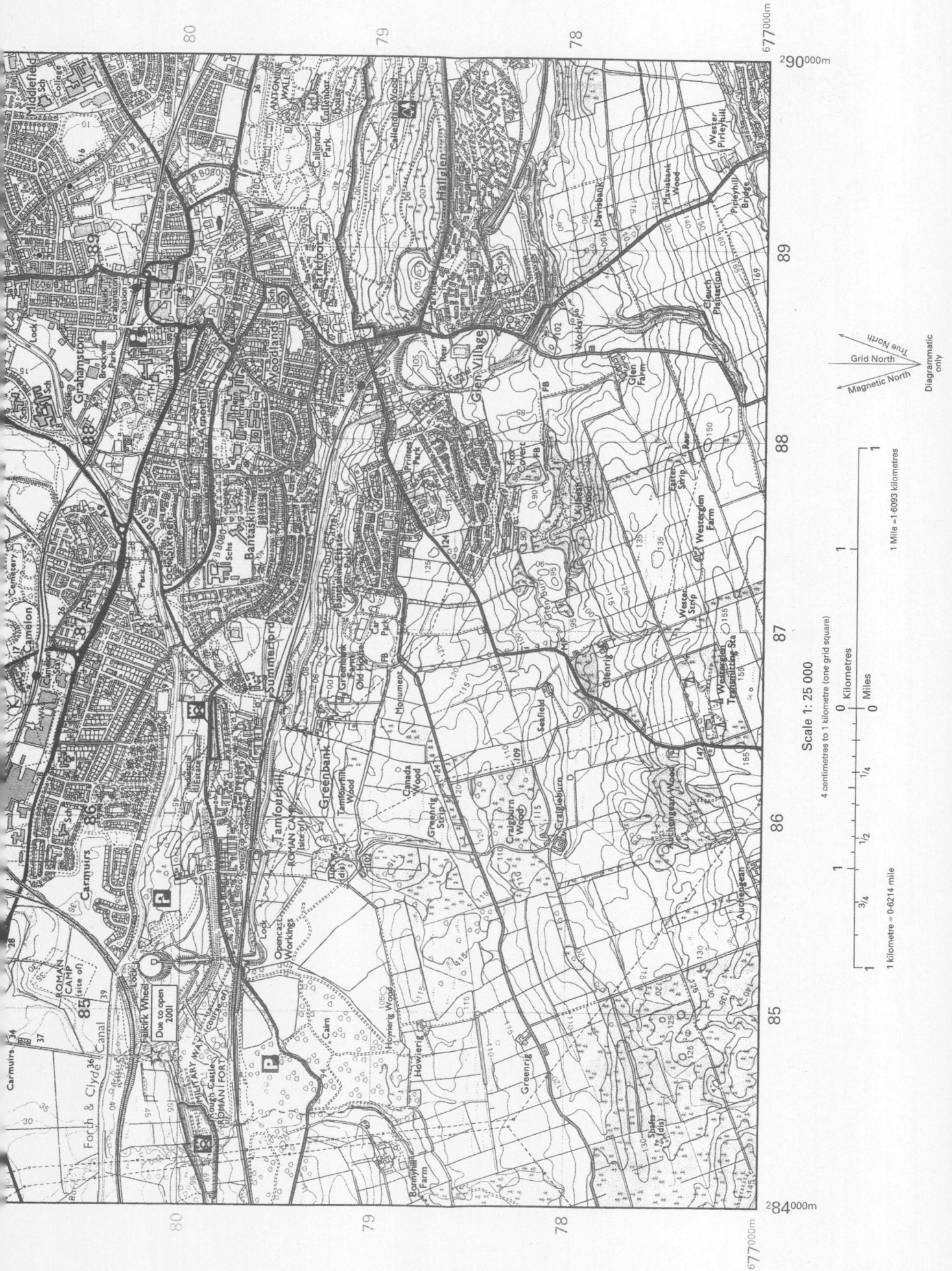

1.

Reference Diagram Q1

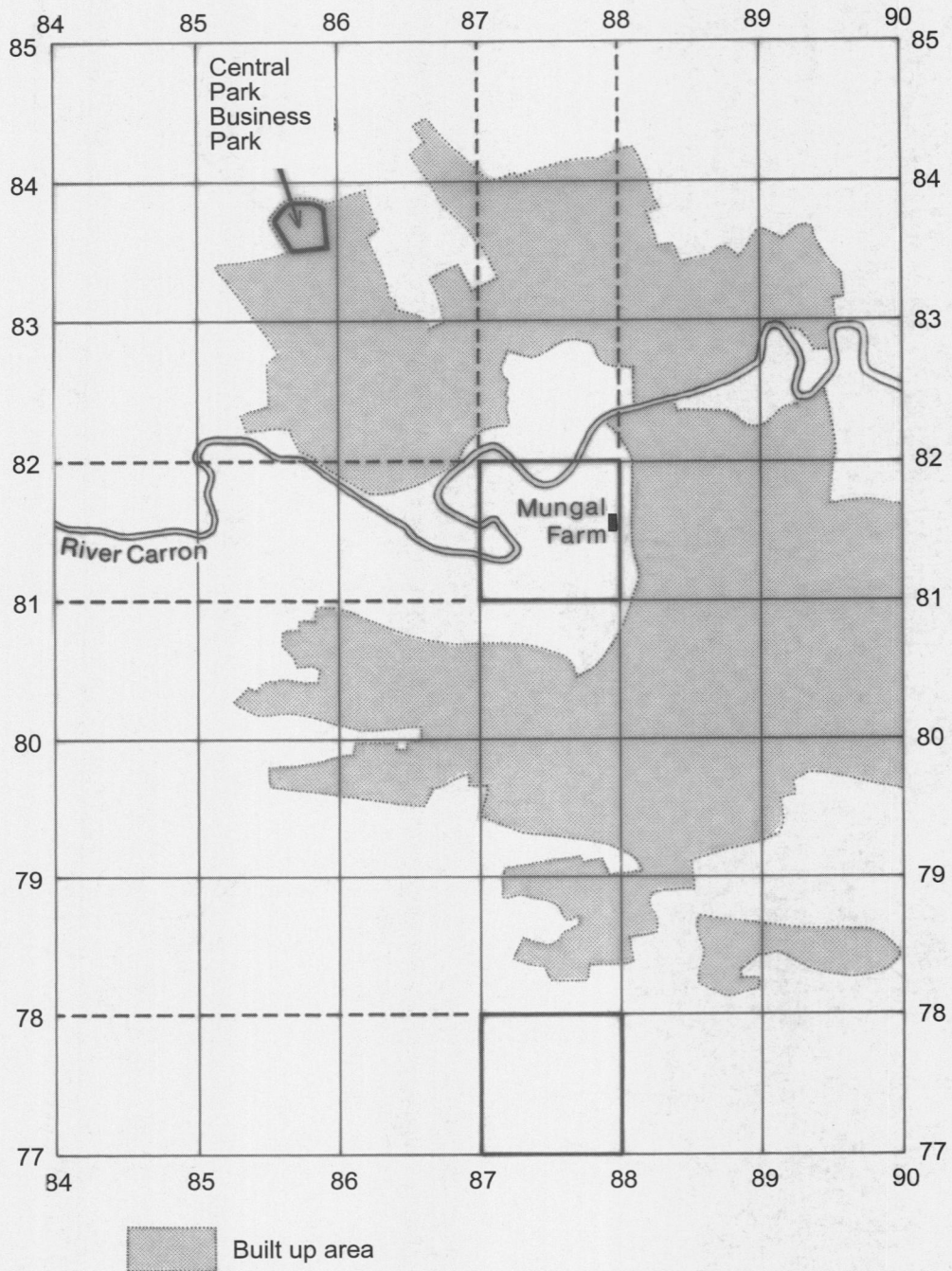

Built up area

Marks

	KU	ES

1. (continued)

This question refers to the OS Map Extract (No 1407/EXP349) of the Falkirk Area and the Reference Diagram Q1 on *Page two*.

(a) Describe the **physical** features of the River Carron **and** its valley from Lochlands (859818) to Carron House (897829).

4

(b) What is the main present day function of Falkirk?

Choose from:

industrial   service centre   tourism and recreation.

Use map evidence to support your answer.

5

(c) Mungal Farm is found at 880815.

Using map evidence, give the advantages **and** disadvantages of the location of this farm.

4

(d) Look at grid squares 8781 and 8777.

Give reasons for the low population density in **each** grid square.

4

(e) Use map evidence to **explain** the location of the Central Park Business Park in grid square 8583.

5

(f) Do you agree that grid square 8880 contains the CBD of Falkirk?

Explain your answer.

4

[Turn over

Mar

KU

2.

Reference Diagram Q2: U-shaped Valley

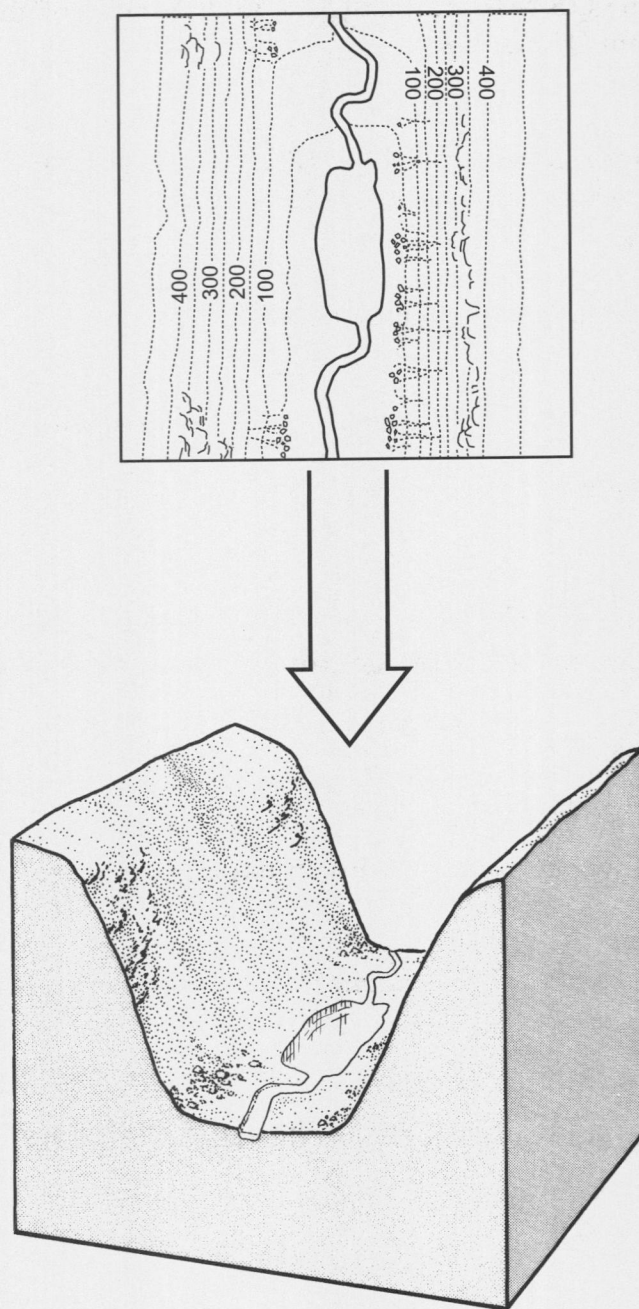

Study Reference Diagram Q2 above.

Explain the formation of a U-shaped valley.

You may use diagrams to illustrate your answer.

4

Marks

KU   ES

3.

Reference Diagram Q3: Land Use Conflict in Loch Lomond and Trossachs National Park

Commercial forestry—active felling in forest

Thousands of walkers on the mountain each year

Traditional hill sheep farming

KILLIN   Loch Tay

A 85

L Voil   A 84   L Earn

Loch Katrine   The Trossachs

CALLANDER

L Venachar

ABERFOYLE

A 811

Loch Lomond

BALMAHA

DRYMEN

10 km

Proposed chalet redevelopment 20 new units

Sites of scientific interest

Cycle paths used by ramblers and mountain bikers

Small residential community

Traffic congestion on main road

Lochs

Land over 600 m

▲ Selected mountains (over 800 m)

Built up area

● Village

Main roads

Boundary of National Park

Study Reference Diagram Q3 above.

(a) Select **two** different land uses.

Explain in detail why they may be in conflict with each other.   5

(b) A group of pupils wants to investigate land use conflicts in the National Park. Describe **two** gathering techniques they could use to collect appropriate data.

Give reasons for your choice.   6

[Turn over

Mar
KU

4. **Reference Diagram Q4A: Weather Conditions in the British Isles**

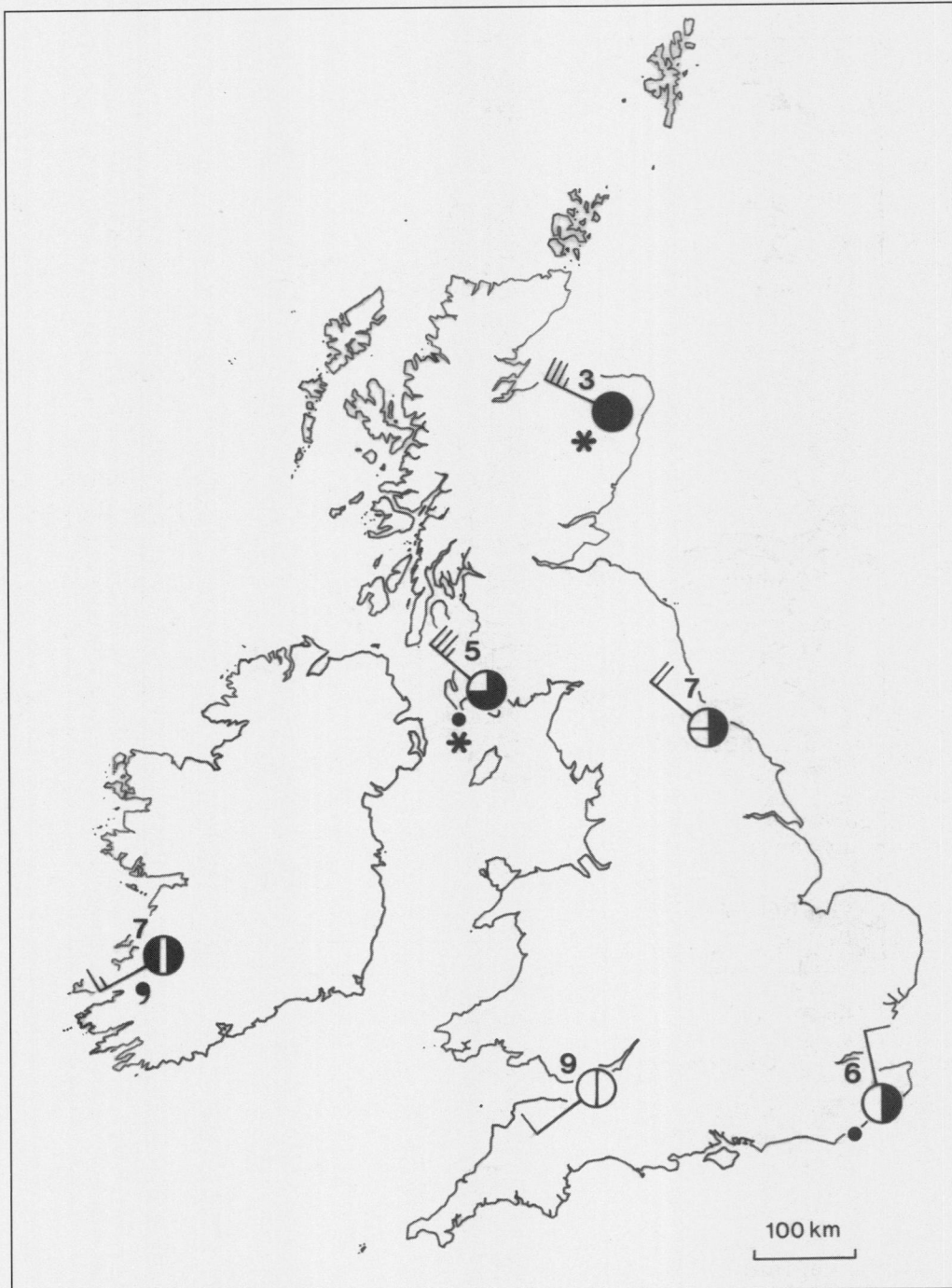

(a) Look at Reference Diagram Q4A.

Compare, in detail, weather conditions in North East Scotland with those in South East England.

4. (continued)

Reference Diagram Q4B: Synoptic Chart for 1200 hrs, 14 November 2004

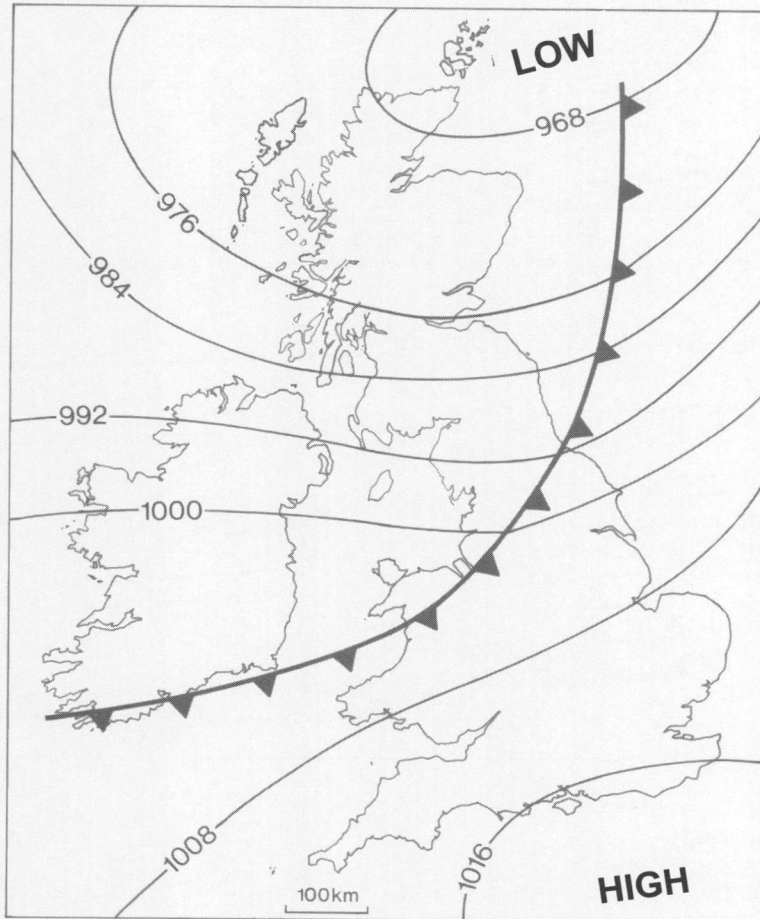

(b) Look at Reference Diagrams Q4A and Q4B.

Do you agree that the weather conditions shown in Reference Diagram Q4A match the synoptic chart in Reference Diagram Q4B?

Explain your answer in detail.

5

[Turn over

Mar
KU

5.   **Reference Diagram Q5: Llanwern Steelworks, South Wales**

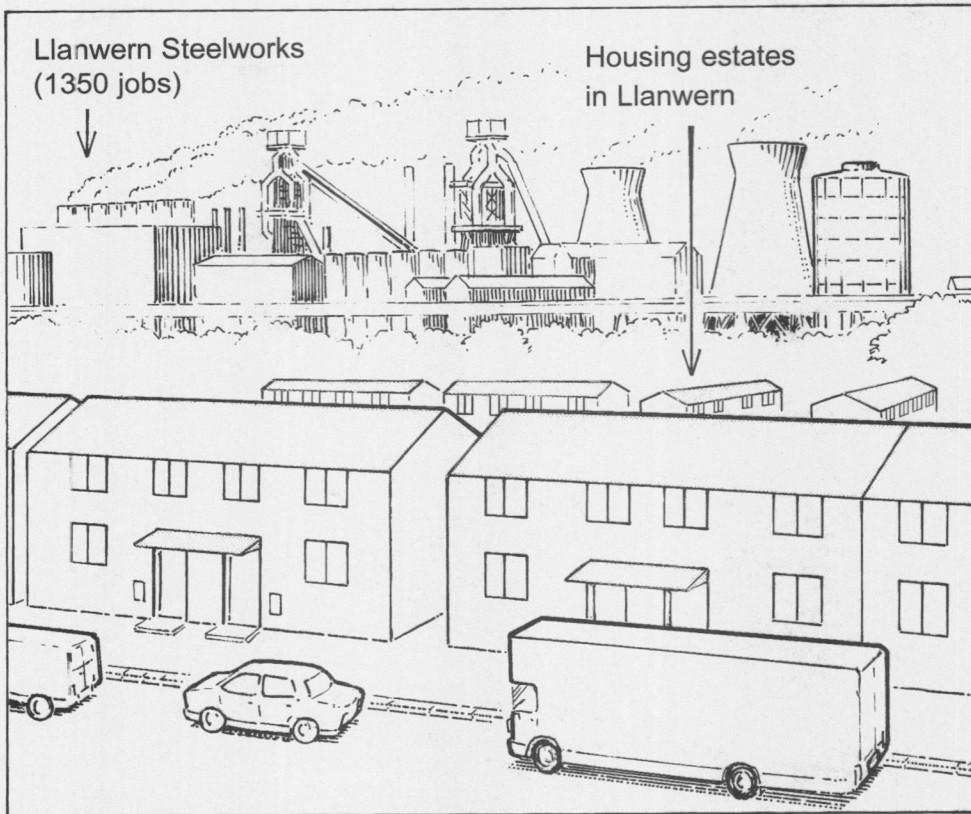

Llanwern Steelworks
(1350 jobs)

Housing estates
in Llanwern

Look at Reference Diagram Q5.

Explain in detail the social, economic and environmental impact of the closure of a large steelworks such as Llanwern on the surrounding area.

5

6. **Reference Diagram Q6A: Changing Farmscape in UK**

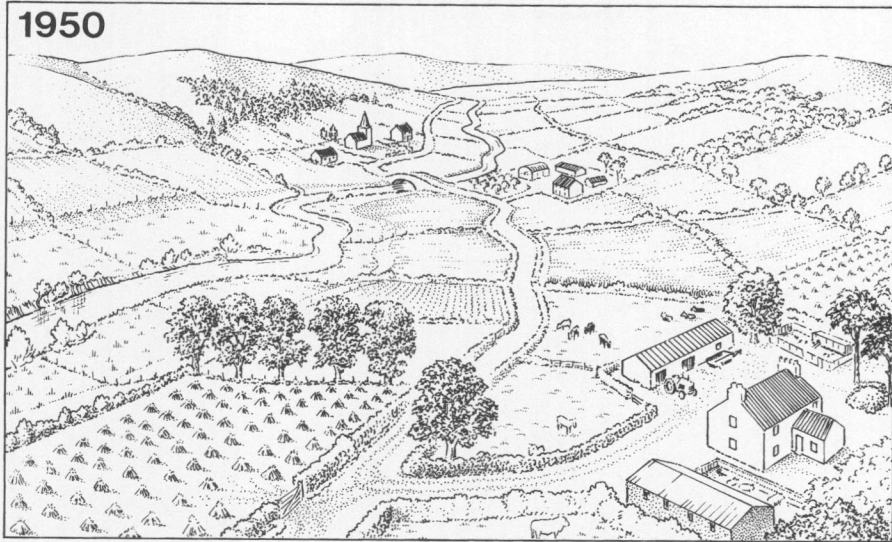

1950

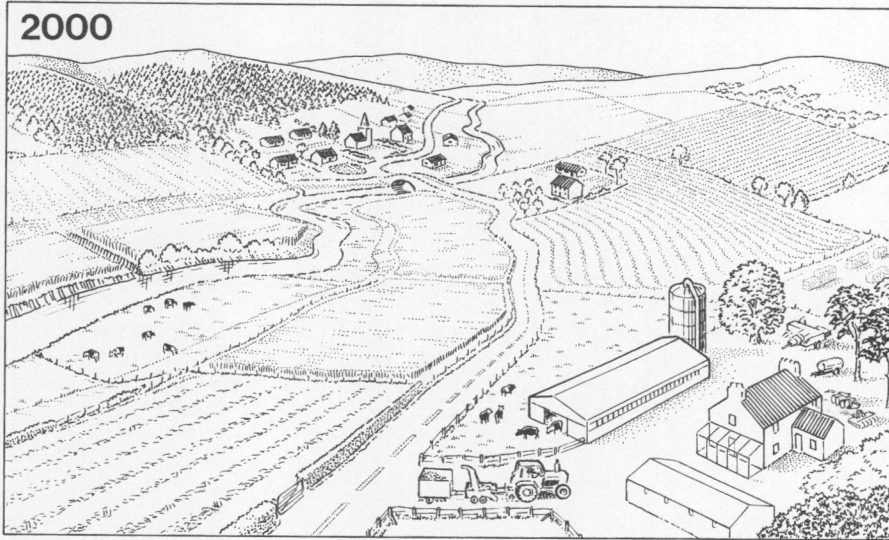

2000

Reference Diagram Q6B: Changes in Farm Size

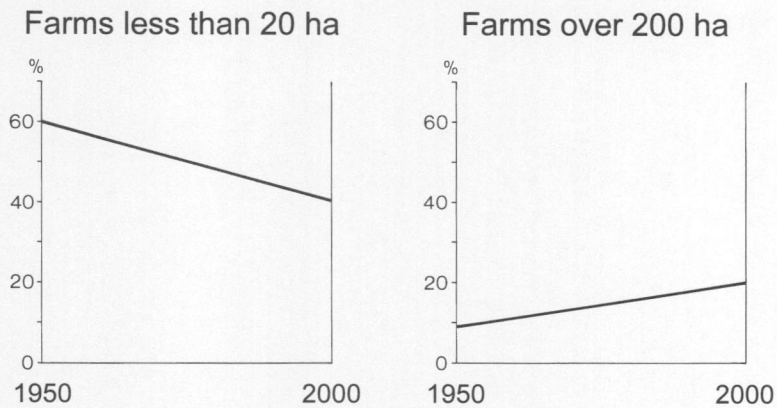

Farms less than 20 ha

Farms over 200 ha

Look at Reference Diagrams Q6A and Q6B.

Explain the advantages **and** disadvantages of the changes shown.

6

Mar

KU

7. **Reference Diagram Q7: Street Maps of two Areas in Greater Glasgow**
(Scale 1:10 000)

Area A

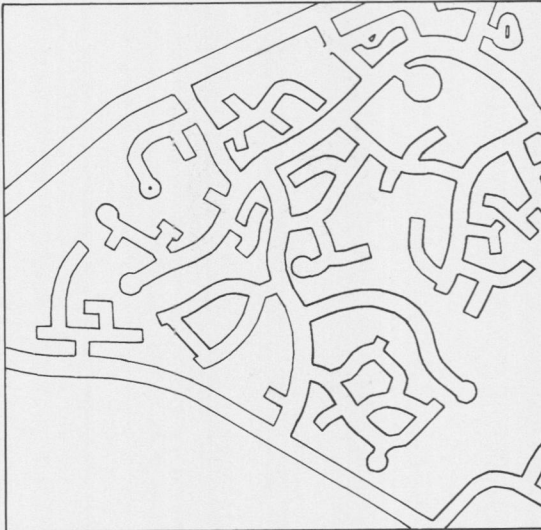

Area B

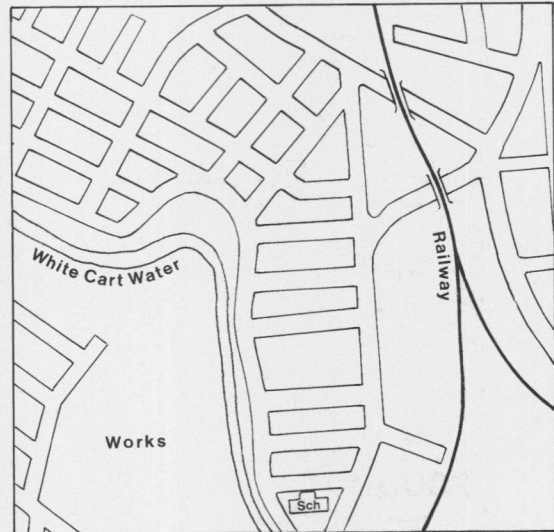

500 metres

Look at Reference Diagram Q7.

Explain why the urban environments in Areas A and B are different.

Your answer may refer to age, quality of environment, street pattern and location.

6

Page ten

8.   **Reference Diagram Q8: Exports of selected Economically Less Developed Countries (Developing Countries)**

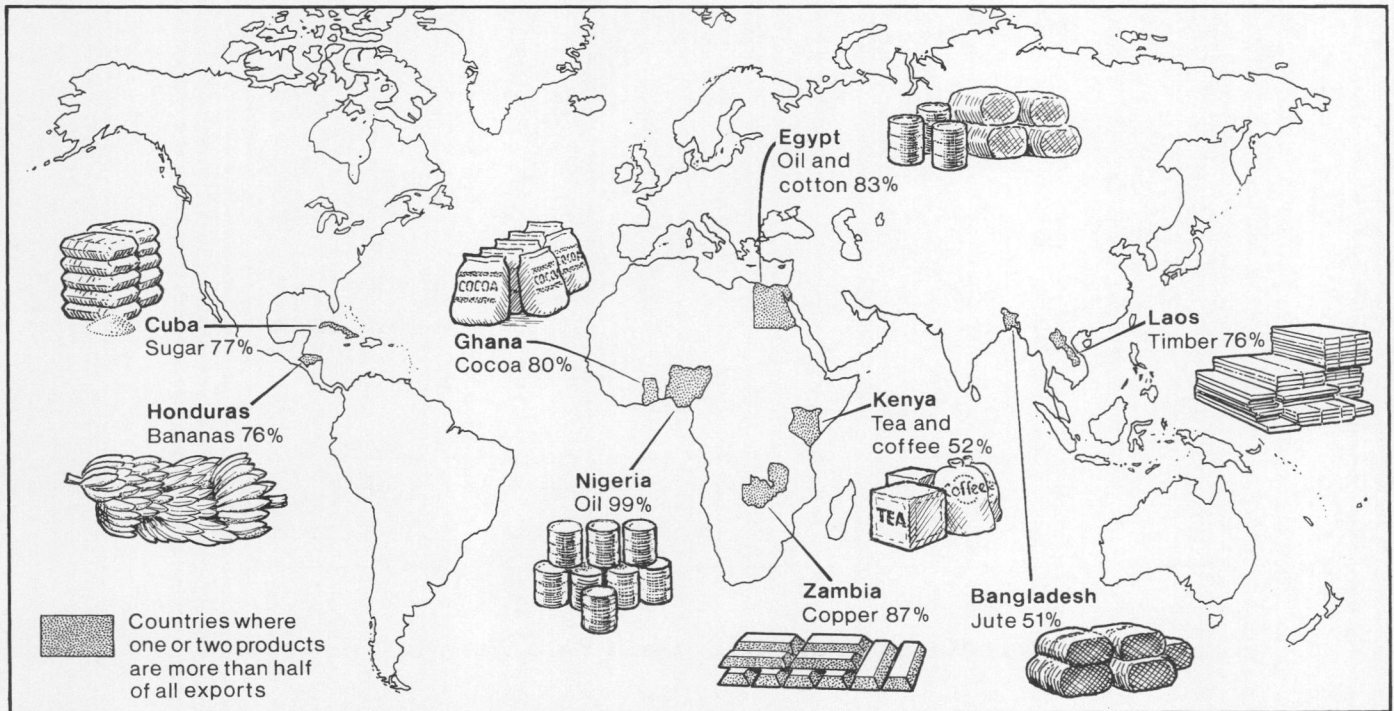

	Marks	
	KU	ES

(a) Look at Reference Diagram Q8.

Explain why some Economically Less Developed Countries (Developing Countries) are especially at risk from changing world prices for the goods which they export.

4

(b) Which other processing techniques could be used to display the export percentage figures shown on Reference Diagram Q8?

Give reasons for your choice of techniques.

6

[Turn over for Question 9 on *Page twelve*

Mark

KU

9.   **Reference Diagram Q9A: Demographic Transition Model**

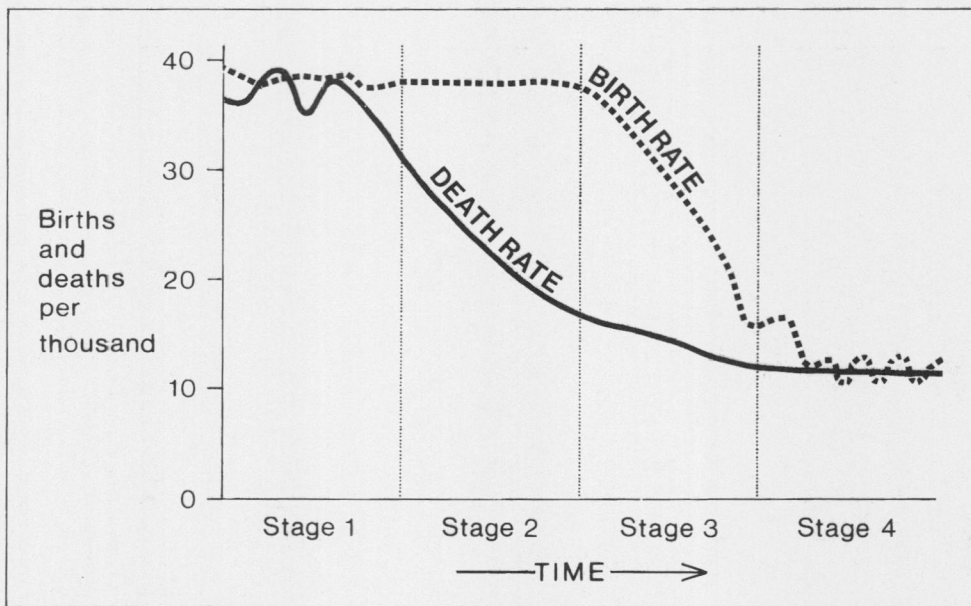

Reference Diagram Q9B: Birth Rate Statistics

Country	Birth Rate/1000
United Kingdom	13
Ethiopia	38

Study Reference Diagrams Q9A and Q9B above.

Many ELDCs* are at Stage 2 in the demographic transition model while EMDCs* are more likely to be at Stage 4.

Give reasons for the differences in birth rates between ELDCs such as Ethiopia and EMDCs such as the UK.

4

* ELDCs = Economically Less Developed Countries
* EMDCs = Economically More Developed Countries

[END OF QUESTION PAPER]

[BLANK PAGE]

FOR OFFICIAL USE

G

KU	ES

Total Marks

1260/403

NATIONAL
QUALIFICATIONS
2006

WEDNESDAY, 10 MAY
10.25 AM–11.50 AM

GEOGRAPHY
STANDARD GRADE
General Level

Fill in these boxes and read what is printed below.

Full name of centre

Town

Forename(s)

Surname

Date of birth
  Day Month Year

Scottish candidate number

Number of seat

1 Read the whole of each question carefully before you answer it.

2 Write in the spaces provided.

3 Where boxes like this ☐ are provided, put a tick ✓ in the box beside the answer you think is correct.

4 Try all the questions.

5 Do not give up the first time you get stuck: you may be able to answer later questions.

6 Extra paper may be obtained from the invigilator, if required.

7 Before leaving the examination room you must give this book to the invigilator. If you do not, you may lose all the marks for this paper.

SCOTTISH
QUALIFICATIONS
AUTHORITY
©

1:50 000 Scale
Landranger Series

Extract No 1488/178

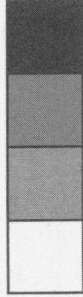

Four colours should appear above; if not then please return to the invigilator.
Four colours should appear above; if not then please return to the invigilator.

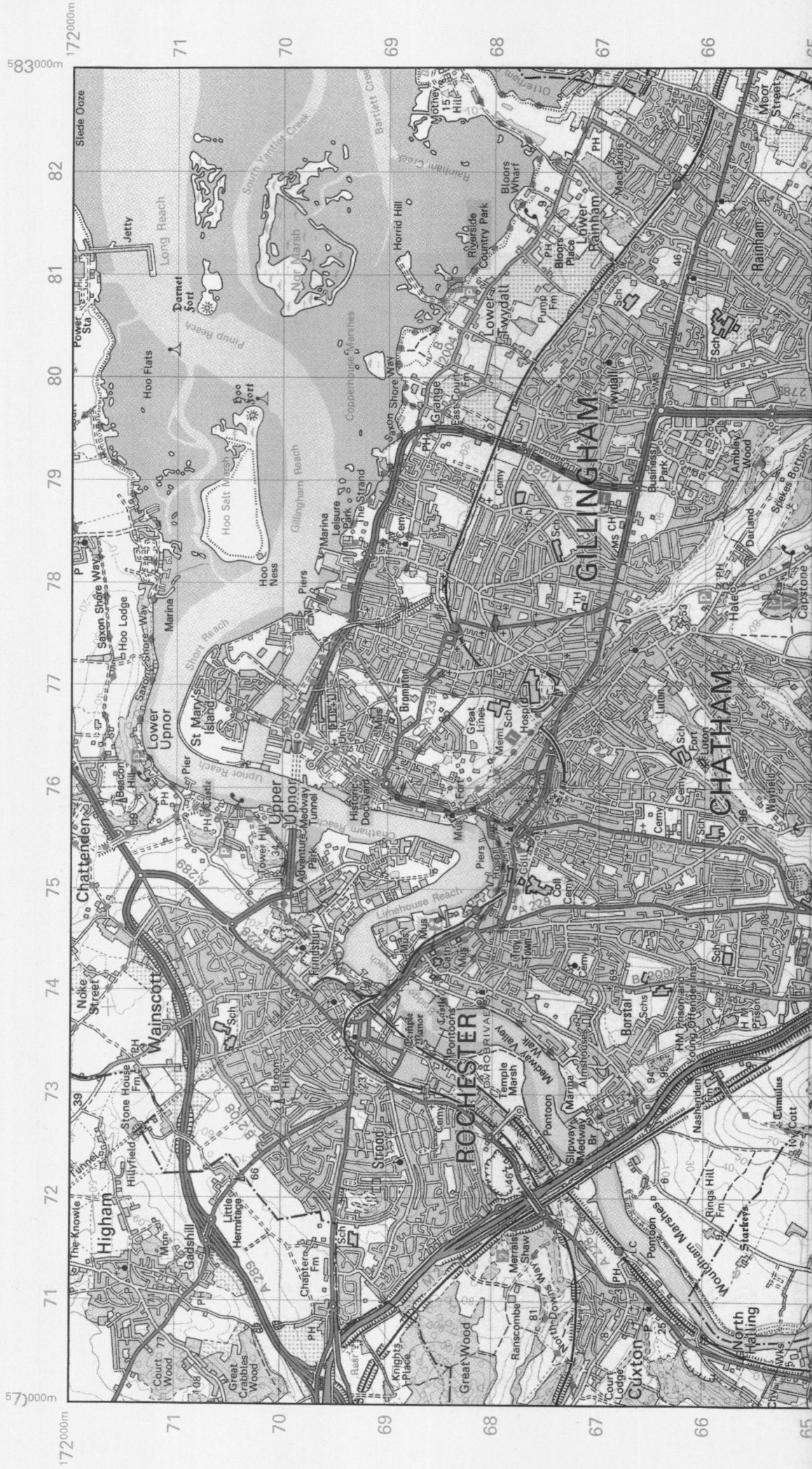

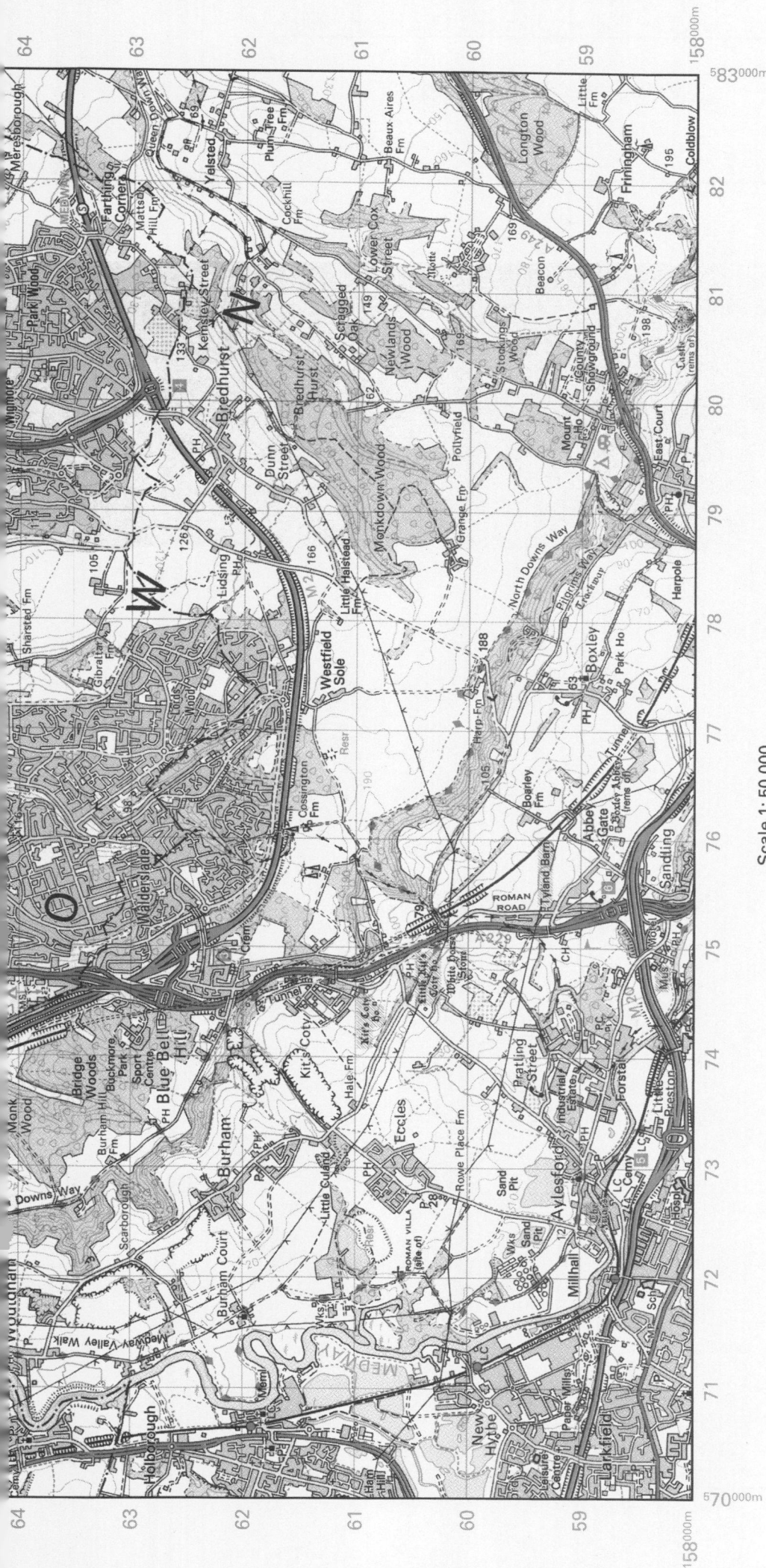

Scale 1:50 000

2 centimetres to 1 kilometre (one grid square)

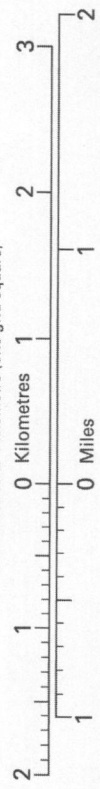

3

2

1 mile = 1·6093 kilometres

1 kilometre = 0·6214 mile

1.

Reference Diagram Q1

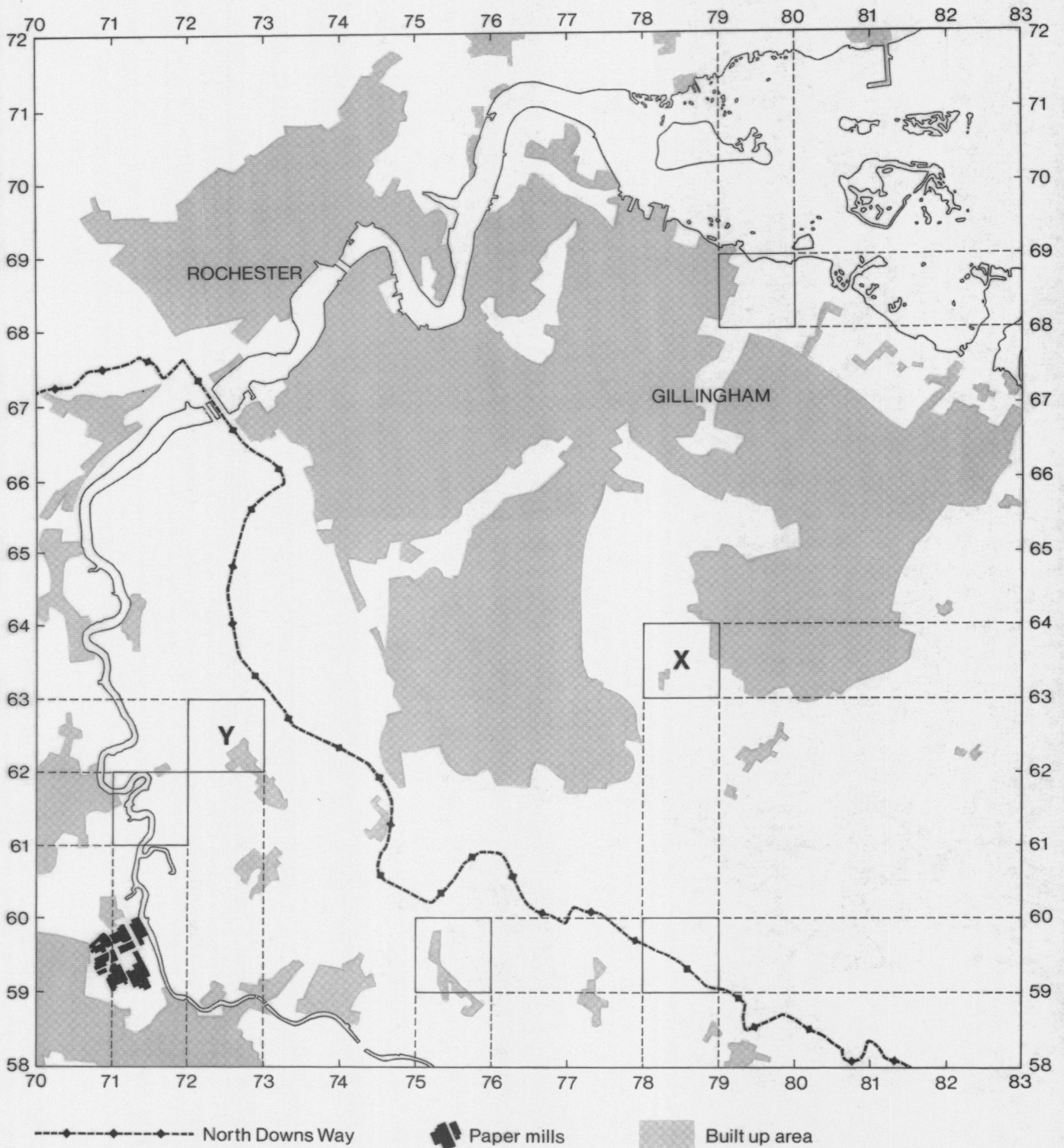

ROCHESTER

GILLINGHAM

X

Y

--◆---◆---◆-- North Downs Way   ■ Paper mills   ▨ Built up area

Marks

1. (continued)

Look at the Ordnance Survey Map Extract (No 1488/178) of the Thames Estuary area and Reference Diagram Q1 on Page two.

(a) **Explain** why an oxbow lake is likely to develop in grid square 7161.

You may use diagrams to illustrate your answer.

The land is flat, The meander bends are very close, on the outside of the bends erosion is happening, river will create a bypass

4

(b) Find the North Downs Way on Reference Diagram Q1 and on the Ordnance Survey Map.

The North Downs Way is a footpath for recreational walkers.

Using map evidence, give the advantages **and** disadvantages of this route.

Good views from height

4

[Turn over

DO NOT
WRITE IN
THIS
MARGIN

KU | ES

Marks

1. (continued)

(c)   Give different reasons why there are trees in the following grid squares.

7559 _____

_____

7859 _____

_____

7968 _____

_____   **3**

(d)   It is proposed to develop either Area X (7863) or Area Y (7262) for new housing.

Which area, X or Y, do you think is more suitable?

Explain your answer **in detail**.

_____

_____

_____

_____

_____

_____   **4**

Marks

1. (continued)

(e)   There is a large paper mill in grid square 7159.

Explain why this site is a suitable location for the paper mill.

You **must** use map evidence.

_____

_____

_____

_____

_____

_____

4

(f)   Rochester and Gillingham are built either side of the River Medway.

In what ways has the River Medway **both** benefited **and** created problems for these settlements?

_____

_____

_____

_____

_____

_____

4

[Turn over

Marks

2.   **Reference Diagram Q2: A Landscape of Glacial Deposition**

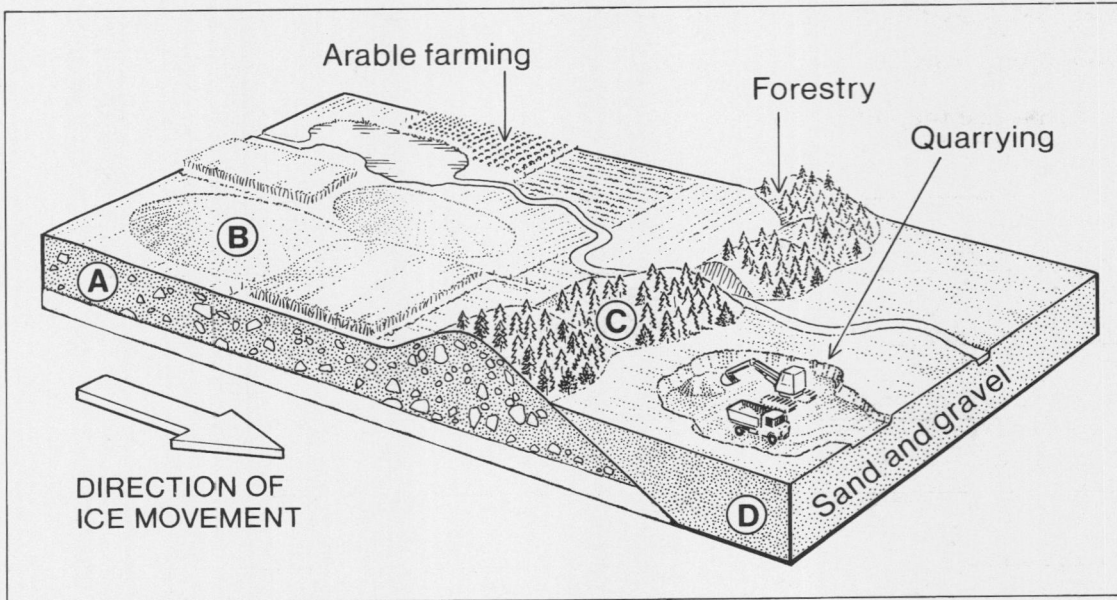

Arable farming

Forestry

Quarrying

B

A

C

D

Sand and gravel

DIRECTION OF
ICE MOVEMENT

(*a*) Match each of the features of glacial deposition in the table to the
correct letter (A, B, C, D) on the Reference Diagram.

Feature	Letter
Drumlin	
Terminal moraine	
Outwash plain	
Boulder clay	

3

DO NOT WRITE IN THIS MARGIN

KU	ES

Marks

2. (continued)

(*b*) **Explain** why **two** of the land uses shown on Reference Diagram Q2 are suitable for the areas in which they are located.

Land Use 1 ______________________

____

____

____

____

Land Use 2 ______________________

____

____

____

______________________ **4**

[Turn over

Marks

3.   **Reference Diagram Q3A: Janice's Radio Phone-in**

Marks

3. (continued)

Reference Diagram Q3B: Synoptic Chart, 10 August 2004

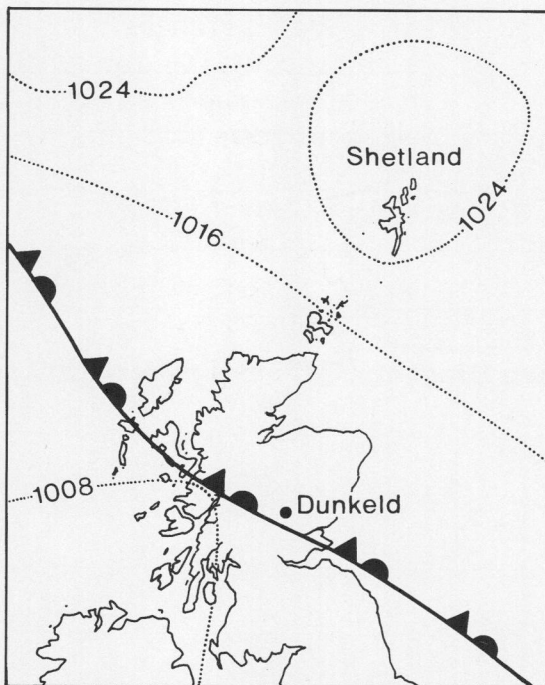

Look at Reference Diagrams Q3A and Q3B.

Explain the different weather experiences which Janice's two callers had on 10 August 2004.

_____

_____

_____

_____

_____

4

[Turn over

DO NOT
WRITE I
THIS
MARGIN

Marks

KU E

4. **Reference Diagram Q4A: Data for six Weather Stations in Canadian Tundra on 3 May 2004**

Station	Hours of Sunshine	Mean Temperature °C
A	2	1
B	6	4
C	8	6
D	3	2
E	5	3
F	2	4

Reference Diagram Q4B: Scattergraph

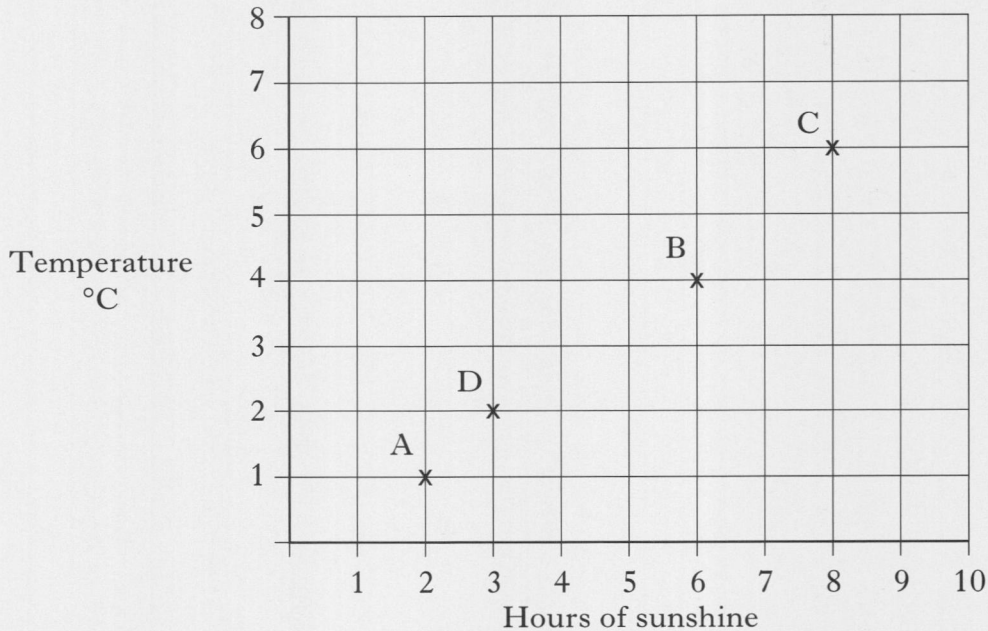

(a) Use the information in Reference Diagram Q4A to complete the scattergraph in Reference Diagram Q4B above.

2

(b) "The scattergraph shows that temperature in the Tundra is directly linked to the hours of sunshine."

Do you agree with this statement?

Give reasons for your answer.

_____

_____

_____

_____

2

Marks

5. **Reference Diagram Q5: Developments in the North Sea**

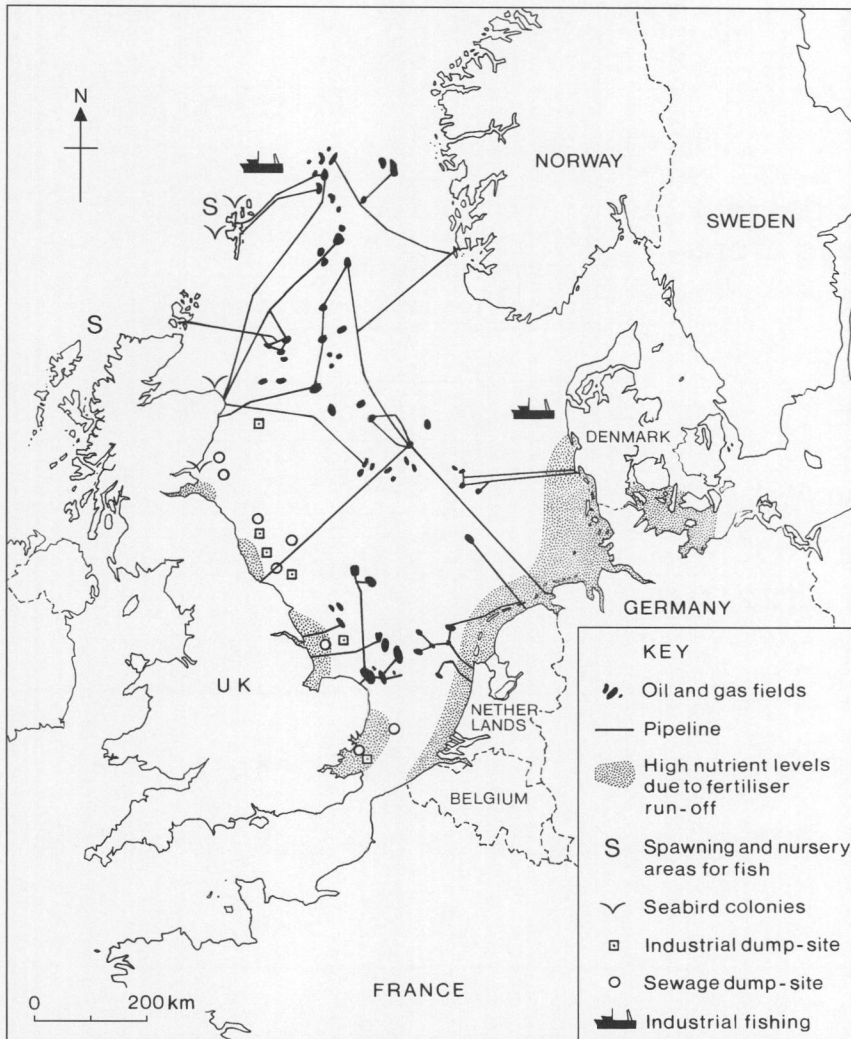

KEY

🖊 Oil and gas fields

— Pipeline

High nutrient levels due to fertiliser run-off

S Spawning and nursery areas for fish

⋎ Seabird colonies

⊡ Industrial dump-site

○ Sewage dump-site

🚢 Industrial fishing

0   200 km

NORWAY

SWEDEN

DENMARK

GERMANY

NETHER-LANDS

BELGIUM

UK

FRANCE

"We must stop exploiting the North Sea. We are destroying its environment."

EU Politician

Do you agree with the statement above?

Give reasons for your answer.

_____

_____

_____

_____

_____

4

6.	**Reference Diagram Q6A:**
	Newspaper Headlines

Reference Diagram Q6B: Examples of
Diversification* on a Farm

SHEEP PRICES
SLUMP

Torrential rain
destroys crops

EU Farm
Subsidies slashed

FOOT AND MOUTH
DISEASE ROCKS
FARMING INDUSTRY

Quad bikes circuit

Holiday cottages

A farm park is a working farm with many rare breeds of farm animals, which visitors pay a small fee to see.

FARM PARK

B & B

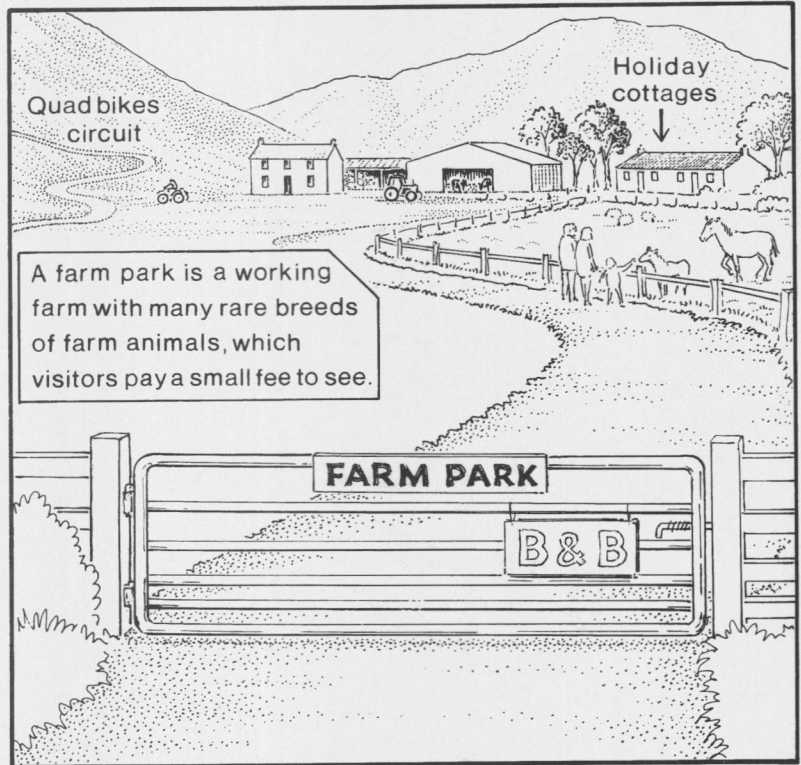

*Diversification = using other land uses to improve farm income

Explain, in some detail, how diversification can help farmers overcome the problems shown in Reference Diagram Q6A.

Marks

4

KU | ES

Marks

7. **Reference Diagram Q7: Factors affecting Location of Industry**

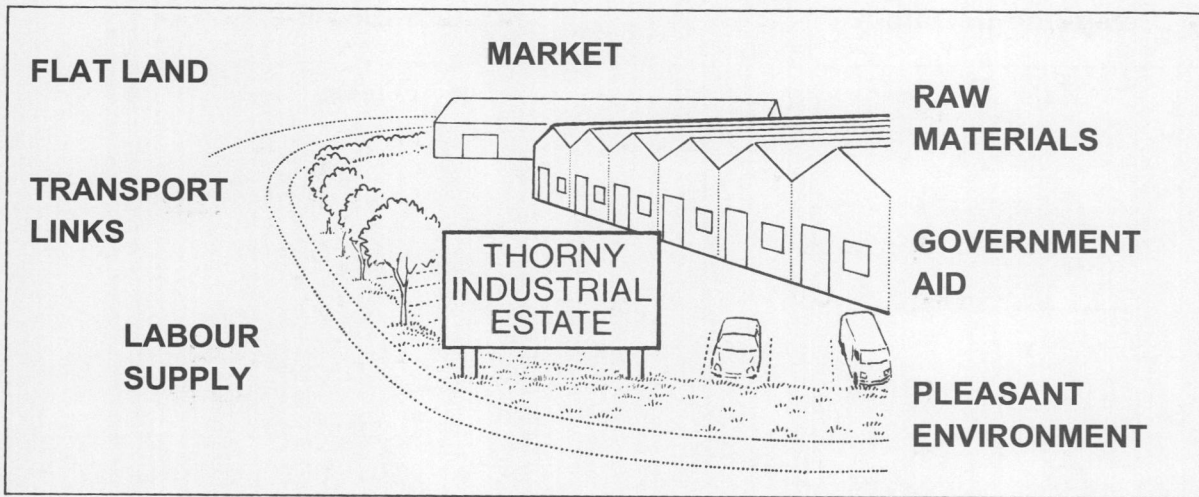

FLAT LAND

MARKET

RAW MATERIALS

TRANSPORT LINKS

THORNY INDUSTRIAL ESTATE

GOVERNMENT AID

LABOUR SUPPLY

PLEASANT ENVIRONMENT

(a) Look at Reference Diagram Q7.

Is Government Aid the most important factor in attracting industry to an area?

Give reasons for your answer.

_____

_____

_____

_____

_____

4

(b) Which techniques could pupils use to gather information about the industrial estate?

Give reasons for your choices.

_____

_____

_____

_____

_____

4

Official SQA Past Papers: General Geography 2006

DO NOT
WRITE IN
THIS
MARGIN

KU | ES

Marks

8. **Reference Diagram Q8: Population Data for two Countries**

Population Data A   **Population Data B**

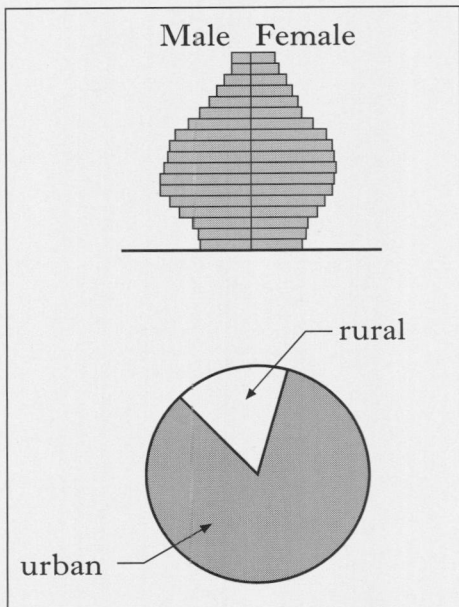

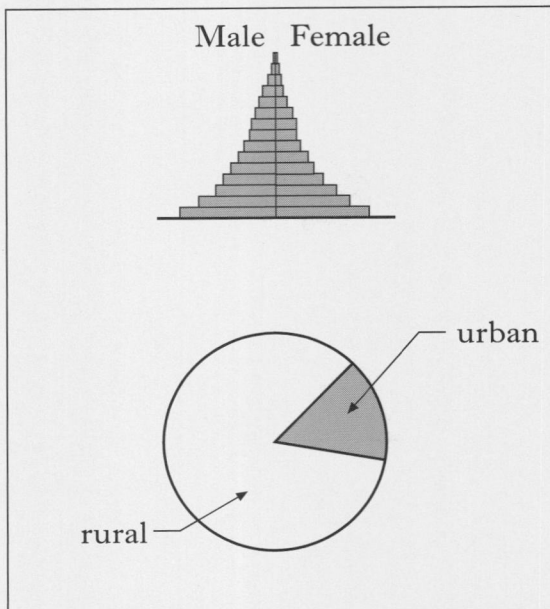

Which set of population data, A or B, is more typical of an Economically Less Developed Country (ELDC)?

Give reasons for your choice.

_____

_____

_____

_____

_____

_____

4

9.　　　　　**Reference Diagram Q9: Japan's Trade Pattern**

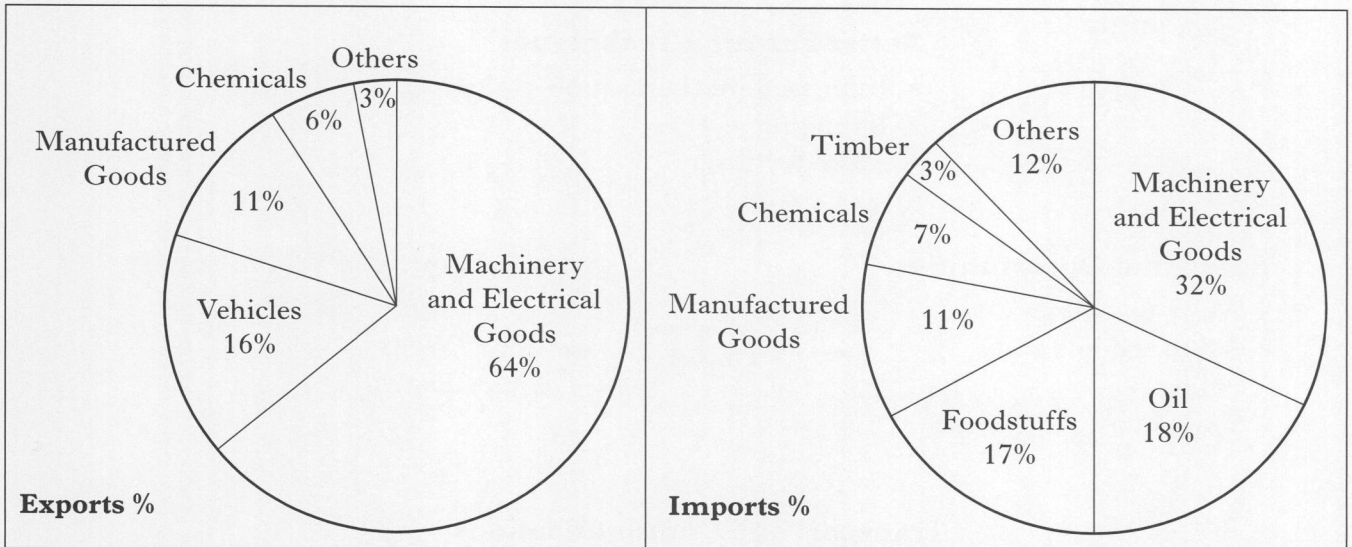

Chemicals

Others
3%

Manufactured
Goods

6%

11%

Vehicles
16%

Machinery
and Electrical
Goods
64%

Exports %

Timber

Others
12%

Chemicals

3%

Machinery
and Electrical
Goods
32%

7%

Manufactured
Goods

11%

Foodstuffs
17%

Oil
18%

Imports %

(a)　Look at Reference Diagram Q9 above.

Do you think this trade pattern is typical of an Economically More Developed Country (EMDC)?

Give reasons for your answer.

Marks

_____

_____

_____

_____

_____

_____

4

(b)　Suggest other processing techniques which could be used to show the information in the pie charts.

Give reasons for your answer.

_____

_____

_____

_____

_____

_____

4

DO NOT
WRITE I
THIS
MARGIN

KU E

Marks

10. **Reference Diagram Q10: Uses of Aid**

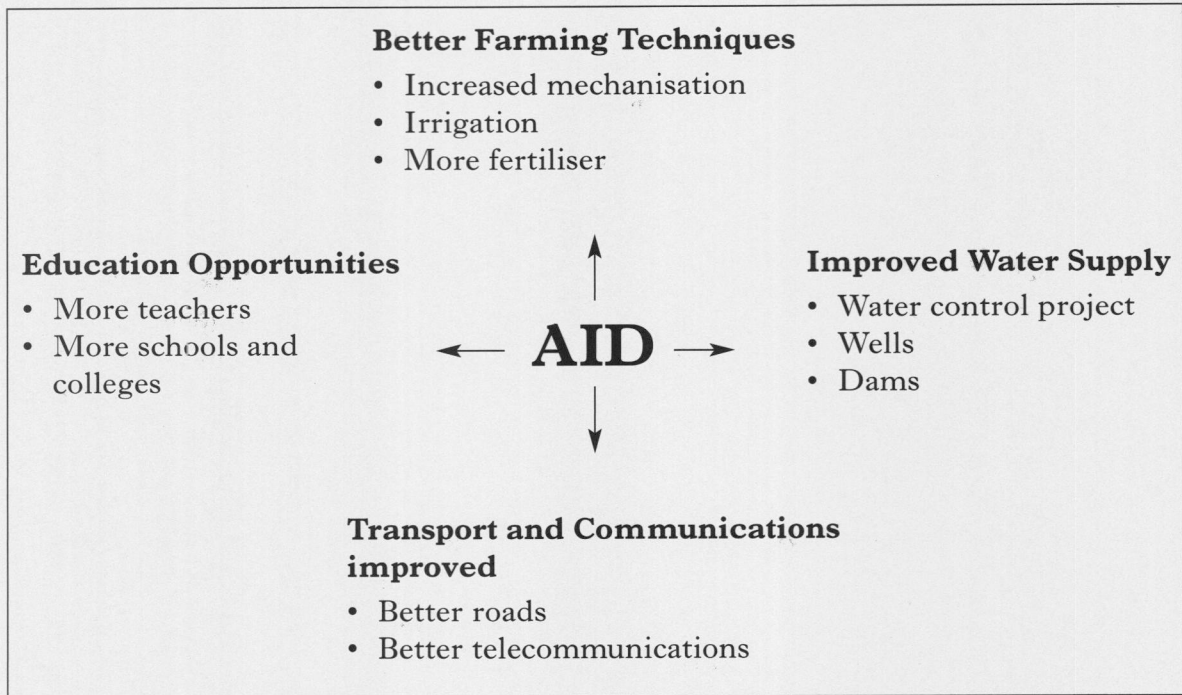

Better Farming Techniques
- Increased mechanisation
- Irrigation
- More fertiliser

Education Opportunities
- More teachers
- More schools and colleges

← **AID** →

Improved Water Supply
- Water control project
- Wells
- Dams

Transport and Communications improved
- Better roads
- Better telecommunications

Which **two** of the above uses of aid do you think would be of **most** benefit to Economically Less Developed Countries (ELDCs)?

Give reasons for your choices.

Choice 1 _____

_____

_____

_____

_____

Choice 2 _____

_____

_____

_____

_____

4

[END OF QUESTION PAPER]

[BLANK PAGE]

C

1260/405

NATIONAL QUALIFICATIONS 2006	WEDNESDAY, 10 MAY 1.00 PM – 3.00 PM	GEOGRAPHY STANDARD GRADE Credit Level

All questions should be attempted.

Candidates should read the questions carefully. Answers should be clearly expressed and relevant.

Credit will always be given for appropriate sketch-maps and diagrams.

Write legibly and neatly, and leave a space of about one cm between the lines.

All maps and diagrams in this paper have been printed in black only: no other colours have been used.

SCOTTISH QUALIFICATIONS AUTHORITY
©

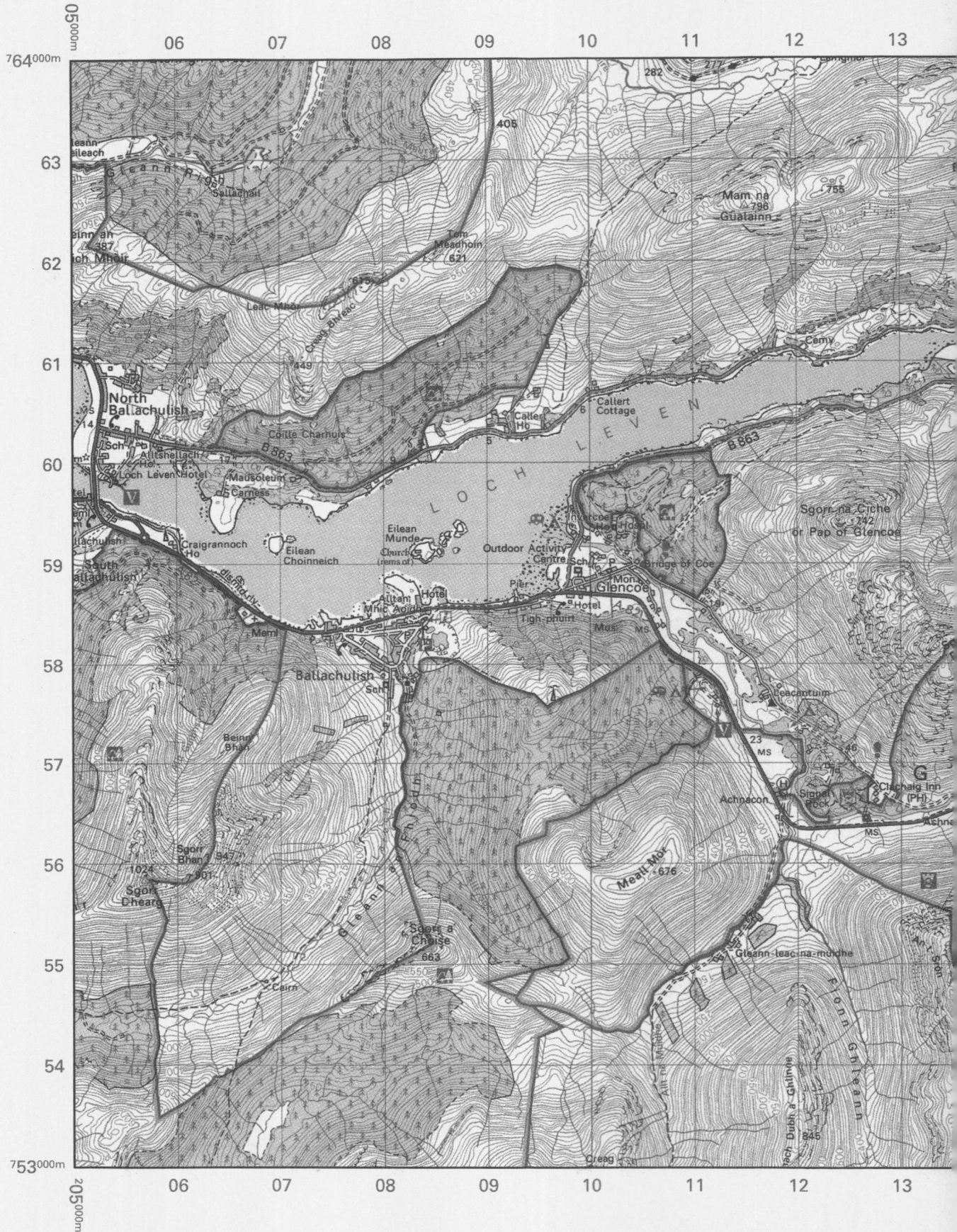

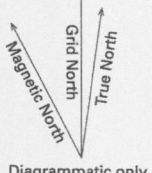

Grid North

True North

Magnetic North

Diagrammatic only

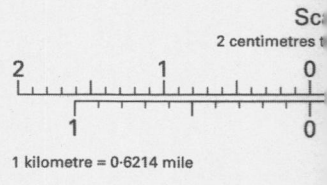

Sc

2 centimetres t

2   0

1   0

1 kilometre = 0·6214 mile

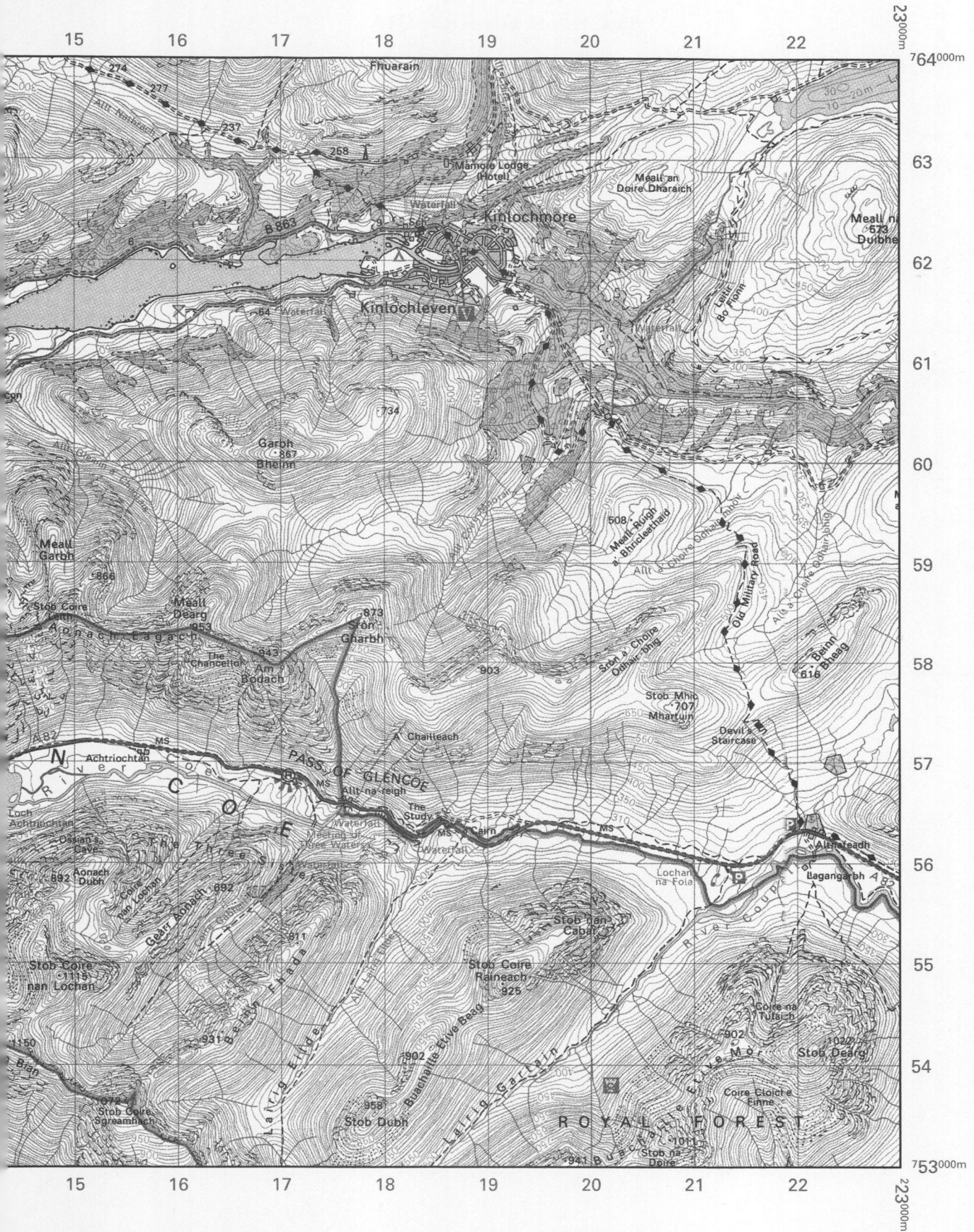

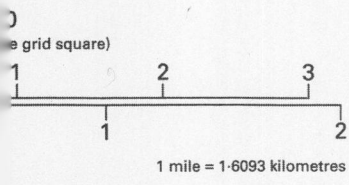

15   16   17   18   19   20   21   22

23 000m

764 000m

Fhuarain

274
277
237
268

Mamore Lodge
(Hotel)

Meall an
Doire Dharaich

B 863

Waterfall

Kinlochmore

Meall na
573
Duibhe

64   Waterfall

Kinlochleven

Waterfall

63

62

61

734

Garbh
867
Bheinn

508
Meall Ruigh
a' Bhricleathaid

Allt a' Choire Odhair-bhig

60

59

Meall
Garbh

866

Stob Coire
Leith

Meall
Dearg
953

873
Stob
Gharbh

Sròn a' Choire
Odhair-bhig

Beinn
Bheag
616

Aonach Eagach

The
Chancellor
943

Am
Bodach

903

Stob Mhic
707
Mhartuin

A82

A' Chailleach

Devil's
Staircase

58

57

N

Achtriochtan

MS

River Coe

PASS OF GLENCOE

Allt-na-reigh

The Study

MS

Old Military Road

MS

56

Loch
Achtriochtan

Ossian's
Cave

C
O
E

The Three Sisters

Waterfall

Meeting of
Three Waters

Waterfall

Cairn

Waterfall

MS

P

Altnafeadh

P

Lagangarbh   A82

Aonach
892
Dubh

892

Coire
nan Lochan

811

Stob nan
Cabar

Lochan
na Fola

55

Stob Coire
1115
nan Lochan

Coire na
Tulaich

931

902

Stob Coire
Raineach
825

Stob Dearg
1022

54

1150

Bidean

1072
Stob Coire
Sgreamhach

958
Stob Dubh

902

Buachaille Etive Beag

Lairig Gartain

Buachaille Etive Mor

R O Y A L   F O R E S T

Coire Cloiche
Finne

1011

Stob na
Doire

53 000m

941

223 000m

15   16   17   18   19   20   21   22

Extract produced by Ordnance Survey 2005
© Crown copyright 2002. All rights reserved.

(e grid square)

1   2   3

1   2

1 mile = 1·6093 kilometres

1.

Reference Diagram Q1A

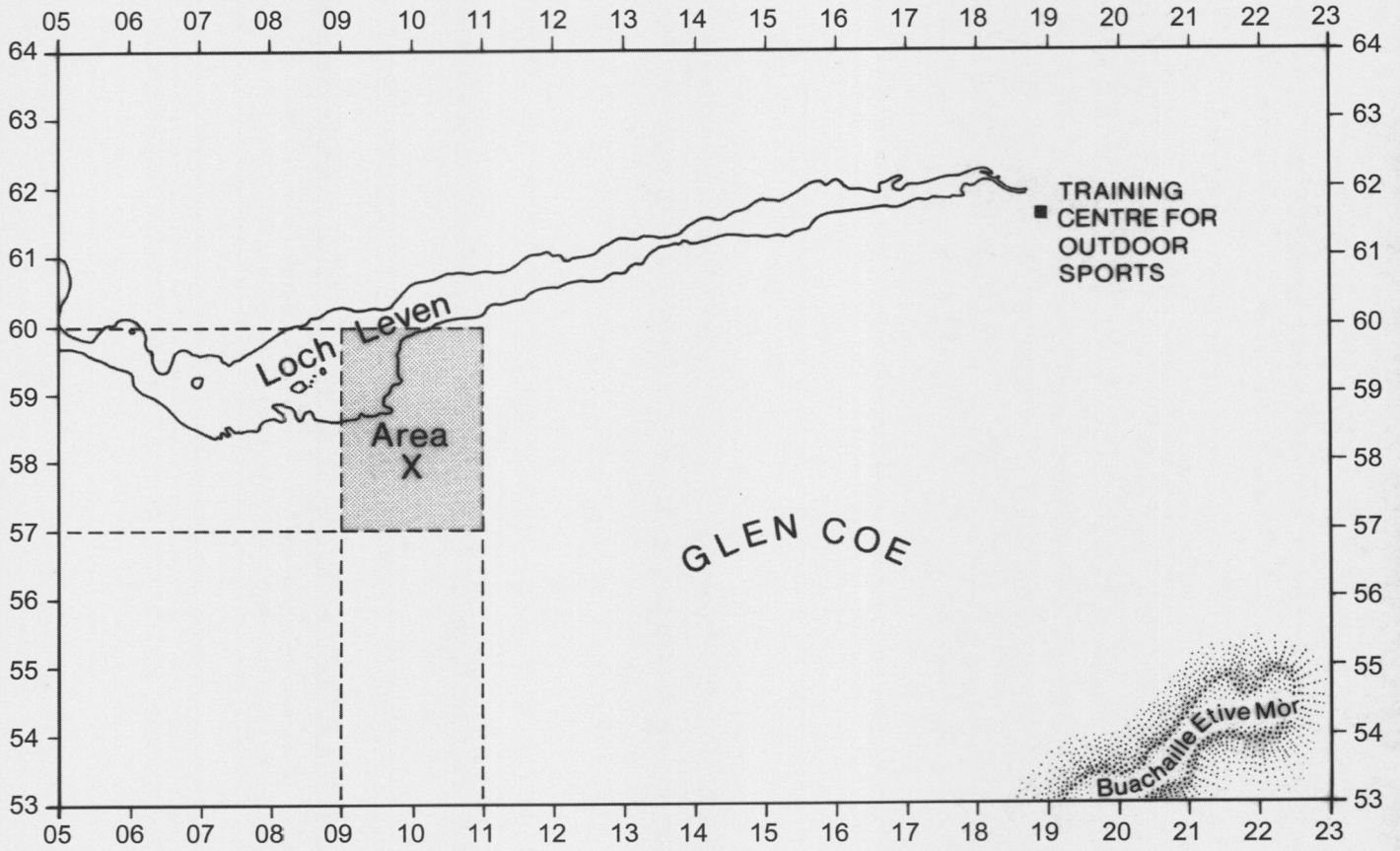

Marks

KU	ES

1. (continued)

This question refers to the OS Map Extract (No 1489/41) of the Glen Coe area and Reference Diagram Q1A on *Page two*.

Reference Diagram Q1B: View looking West from 197618

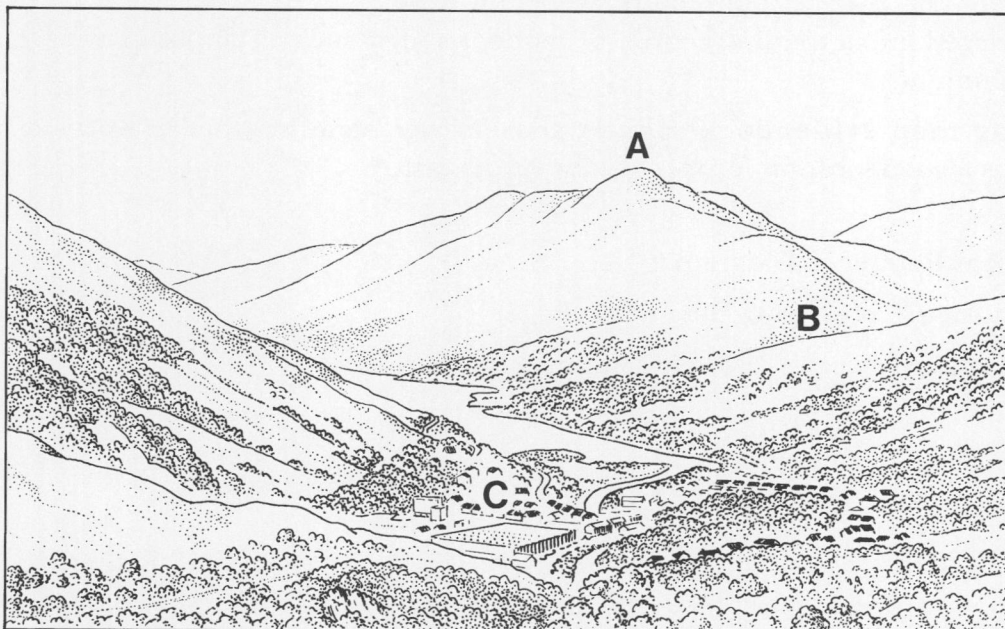

(a) Look at Reference Diagram Q1B and the map extract.

Reference Diagram Q1B is a view looking west from 197618.

Identify the **three** features marked A, B and C.

Choose from:

Allt Nathrach;   Kinlochleven;   Mam na Gualainn;   Kinlochmore;

Beinn na Caillich.

3

(b) (i) Match each of the features named below with the correct grid reference.

Features: **arete;   hanging valley;   truncated spur;   corrie**.

Choose from grid references: 165553,   057563,   197584,   201556.

3

(ii) **Explain** how **one** of the features listed in (b)(i) was formed.

You may use diagrams to illustrate your answer.

4

[Turn over

Ma
KU

1. (continued)

(c) "Glen Coe and the surrounding area is one of Scotland's most popular tourist areas."

Spokesperson for the Scottish Tourist Industry

Part of the disused aluminium works at Kinlochleven (1861, 1862) has been converted into a training centre for outdoor sports such as climbing, walking and ice climbing.

Using map evidence to support your answer, state whether or not you think this is a good location for an outdoor sports centre.

(d) Look at Reference Diagram Q1A.

Find Area X on the OS map extract.

Give reasons for the different land uses in this area.

5

Reference Diagram Q1C: A Wind Farm

(e) A developer is proposing to build a wind farm of up to thirty wind turbines on Buachaille Etive Mòr (see Reference Diagrams Q1A and Q1C).

Do you think this proposal should go ahead?

You **must** use map evidence to support your answer.

2.   **Reference Diagram Q2: Synoptic Chart for 15 January 1995**

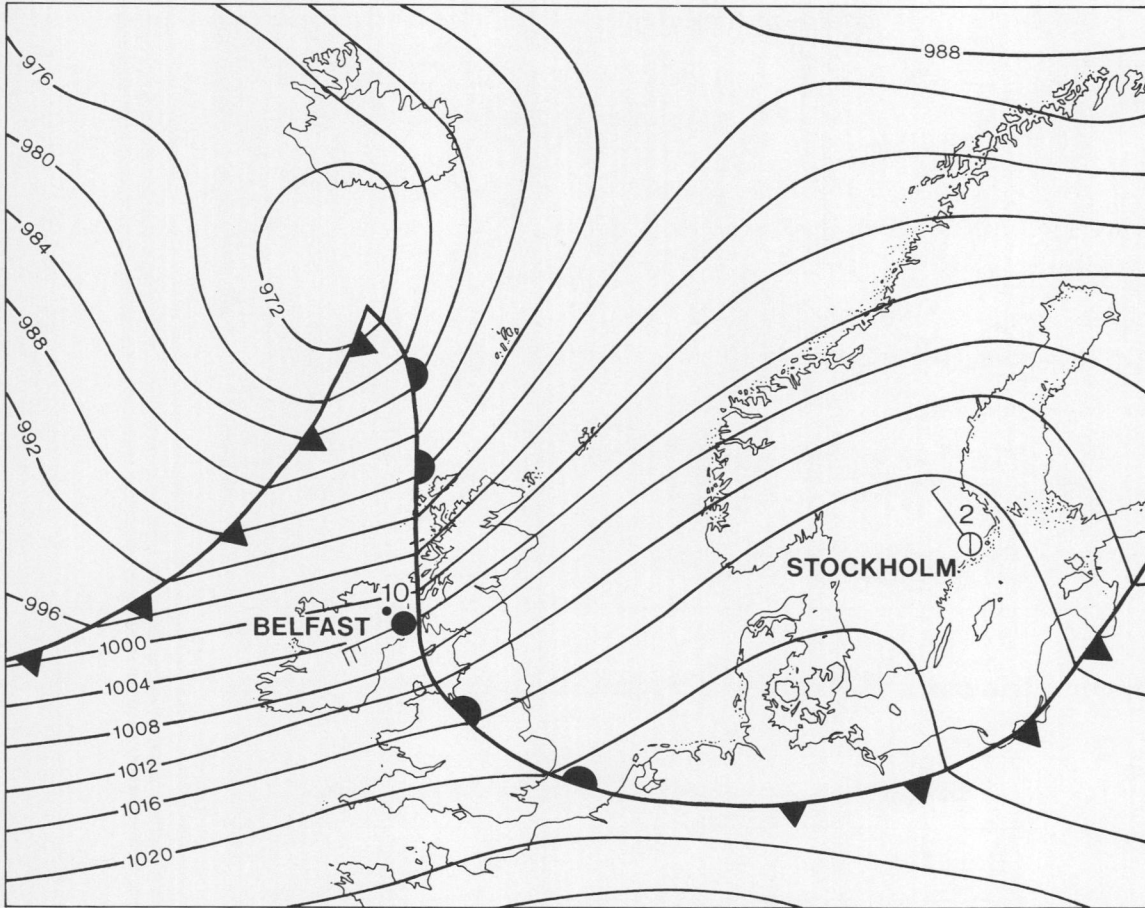

Belfast

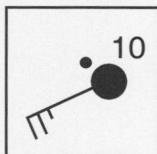

Stockholm

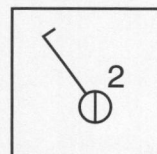

Look at Reference Diagram Q2.

Explain the **differences** in the weather conditions between Belfast and Stockholm.

6

[Turn over

Ma

KU

3.

Reference Diagram Q3A: Destruction of Rainforest

Reference Diagram Q3B: Loss of Rainforest per Year in Selected Countries

Country	Loss of Rainforest per Year
Brazil	6%
Indonesia	10%
Venezuela	12%

Look at Reference Diagrams Q3A and Q3B.

Explain why the world's rainforests continue to be destroyed.

5

Marks

KU   ES

4.   **Reference Diagram Q4A: Recent Changes in Farming**

- Farmers paid to "set aside" land
- Farmers converting to produce organic food
- Removal of hedges to make fields larger
- Increased mechanisation

(a) Study Reference Diagram Q4A.

Do you think these changes **benefit** the countryside?

Give reasons for your answer.

6

Reference Diagram Q4B: Land Use Data for Crow Farm East Anglia

Land Use	Area (hectares)
Barley	13·5
Wheat	12·5
Farm yard and buildings	1·2
Sugar beet	12·0
Vegetables	5·5
Set aside	8·5

(b) Study Reference Diagram Q4B.

Give **two** other techniques which would be appropriate to process this data.

Explain your choices.

5

[Turn over

5.

Reference Diagram Q5A: Land Values in a City

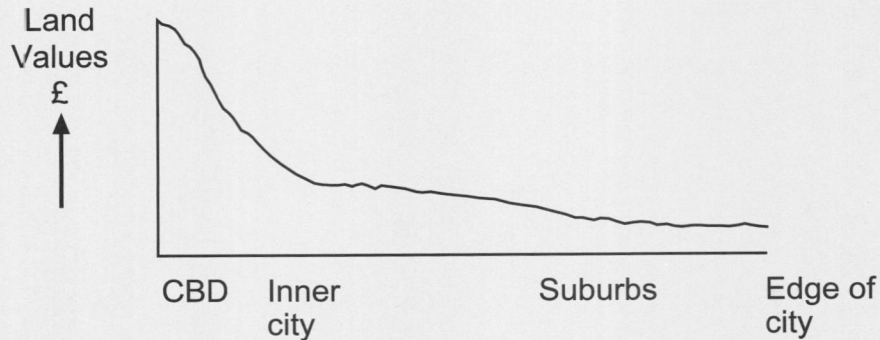

Land
Values
£

CBD   Inner   Suburbs   Edge of
       city   city

Reference Diagram Q5B: Features of Housing Areas

Inner City	Suburbs
Nearer CBD	**Nearer edge of city**
In grid-iron pattern with long, straight streets	Varied street pattern with many short streets in a cul-de-sac arrangement
Houses close to industry	Housing separate from industry
Tenements and/or terraces	Variety of house types, with many detached and semi-detached
Little open space/greenery	More spacious with many gardens
Environmental problems	Pleasant environment

Marks
| KU | ES |

5. (continued)

Look at Reference Diagrams Q5A and Q5B.

(a) **Explain** the features of **either** the Inner City **or** the Suburbs shown in Reference Diagram Q5B.

6

(b) Describe techniques which could be used to gather information about differences between the environments of two urban areas.

Give reasons for your choice of techniques.

5

[Turn over

6. **Reference Diagram Q6: Population Growth in Tokyo and Jakarta**

Tokyo   **Jakarta**

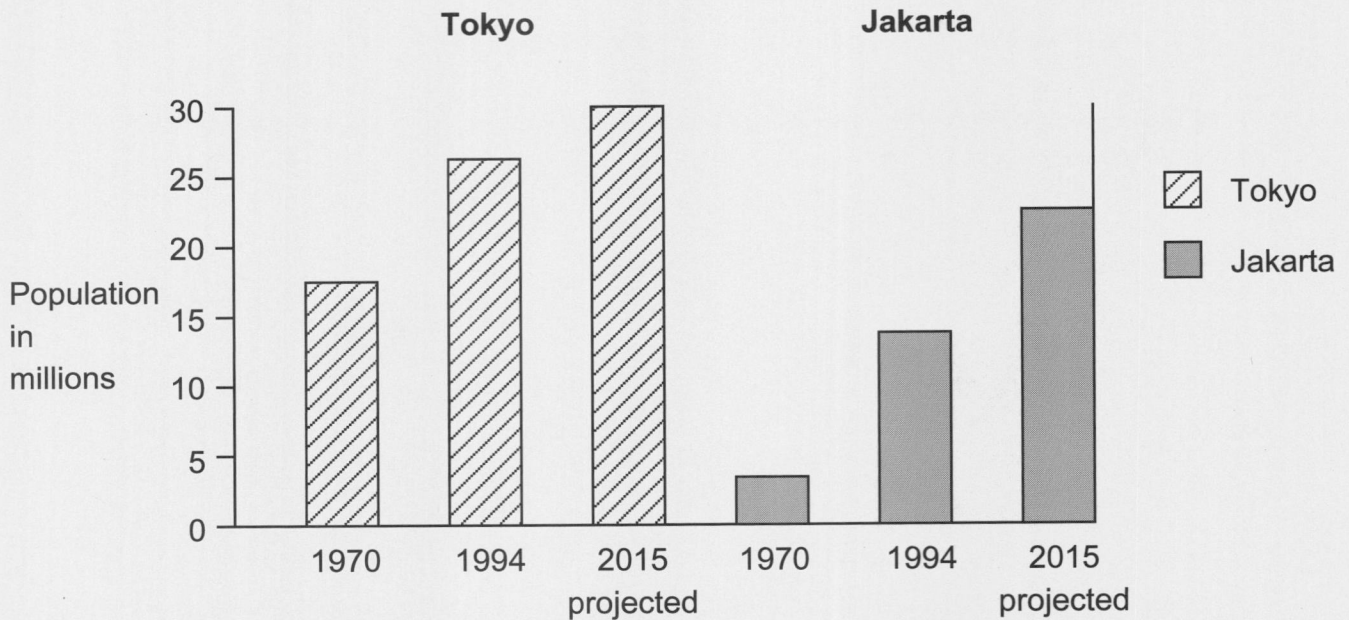

(a) Compare **in detail** the population growth in the two cities.

Tokyo is the capital of Japan, an Economically More Developed Country (EMDC).

Jakarta is the capital of Indonesia, an Economically Less Developed Country (ELDC).

(b) This population growth will cause more problems for Tokyo than for Jakarta.

Do you agree?

Give reasons for your answer.

Mar

KU

Marks

KU | ES

7.   **Reference Diagram Q7: Factors affecting Population Distribution**

relief   accessibility   wealth

natural
resources   **Factors affecting
population distribution**   employment
opportunities

soil   government decisions   climate

Which factors, physical **or** human, have the greater influence on population distribution?
Give reasons for your choice.

6

[Turn over for Question 8 on *Page twelve*

8.   **Reference Diagram Q8: Location of the Three Gorges Dam**

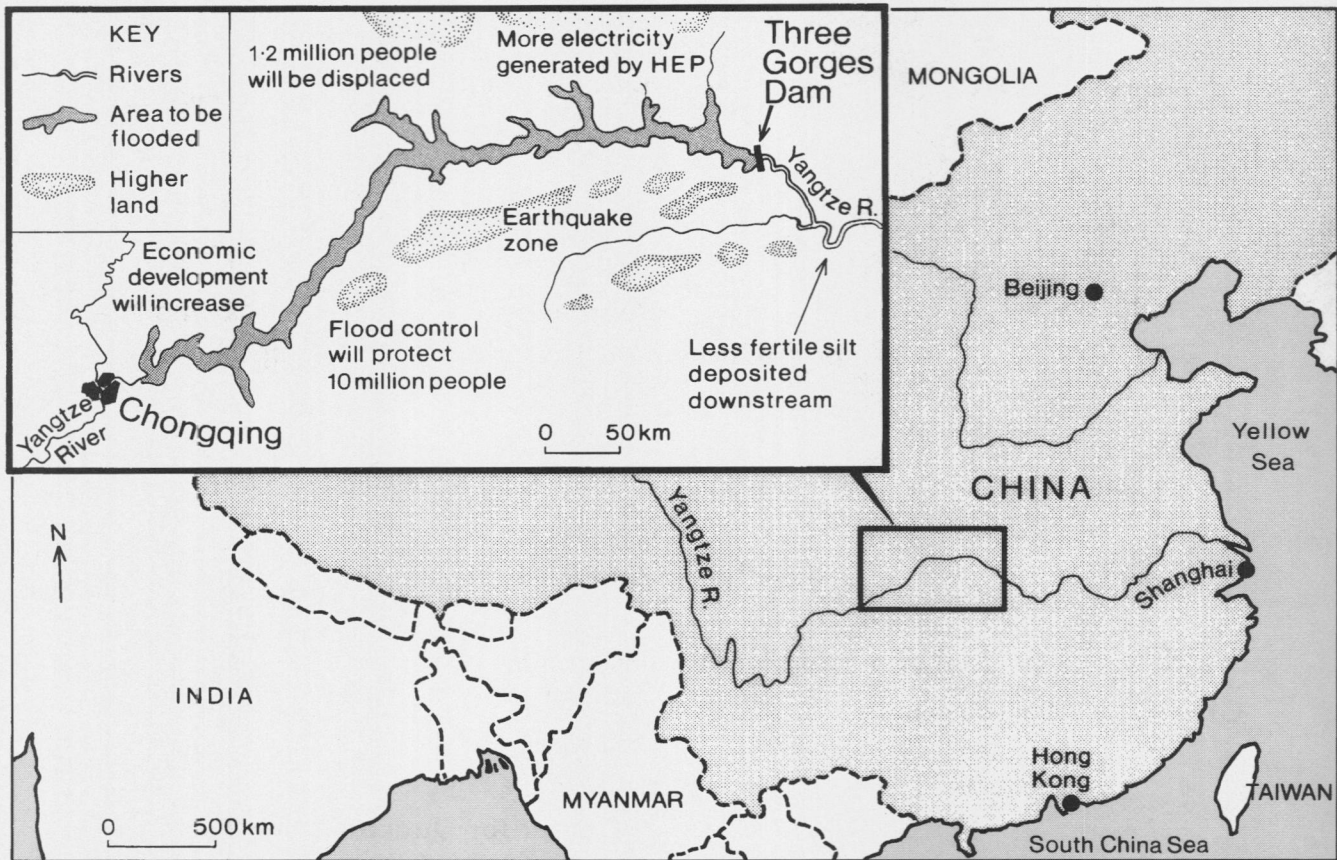

The Three Gorges Dam project was built with money from China and investments from Japan, Canada, Germany and Switzerland. These investments were made in order to develop trade links with China.

What are the advantages **and** disadvantages for China and its trading partners of using these investments to build the Three Gorges Dam?

[END OF QUESTION PAPER]

[BLANK PAGE]

FOR OFFICIAL USE

G

KU | ES

Total Marks

1260/403

NATIONAL
QUALIFICATIONS
2007

TUESDAY, 8 MAY
10.25 AM–11.50 AM

GEOGRAPHY
STANDARD GRADE
General Level

Fill in these boxes and read what is printed below.

Full name of centre

Town

Forename(s)

Surname

Date of birth
Day Month Year

Scottish candidate number

Number of seat

1 Read the whole of each question carefully before you answer it.

2 Write in the spaces provided.

3 Where boxes like this ☐ are provided, put a tick ✓ in the box beside the answer you think is correct.

4 Try all the questions.

5 Do not give up the first time you get stuck: you may be able to answer later questions.

6 Extra paper may be obtained from the invigilator, if required.

7 Before leaving the examination room you must give this book to the invigilator. If you do not, you may lose all the marks for this paper.

SCOTTISH
QUALIFICATIONS
AUTHORITY

©

1:50 000 Scale
Landranger Series

Four colours should appear abo
Four colours should appear abo

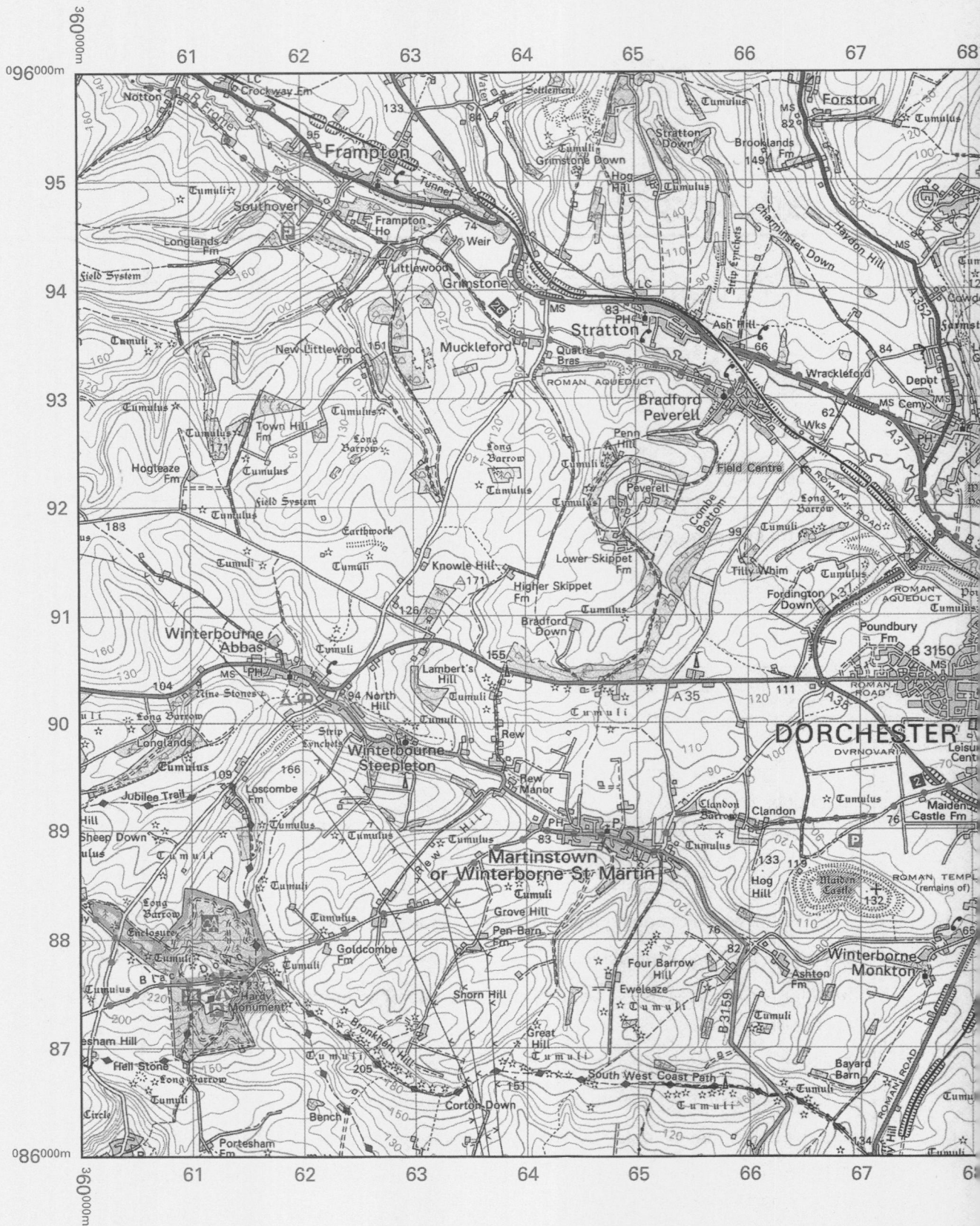

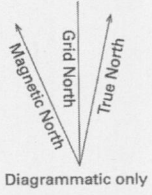

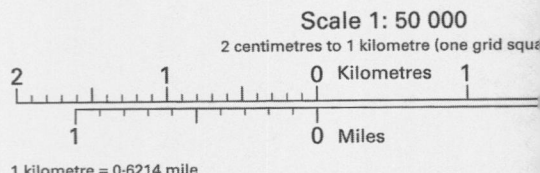

Grid North
True North
Magnetic North

Diagrammatic only

Scale 1: 50 000
2 centimetres to 1 kilometre (one grid squa

2   1   0 Kilometres 1

1   0 Miles

1 kilometre = 0·6214 mile

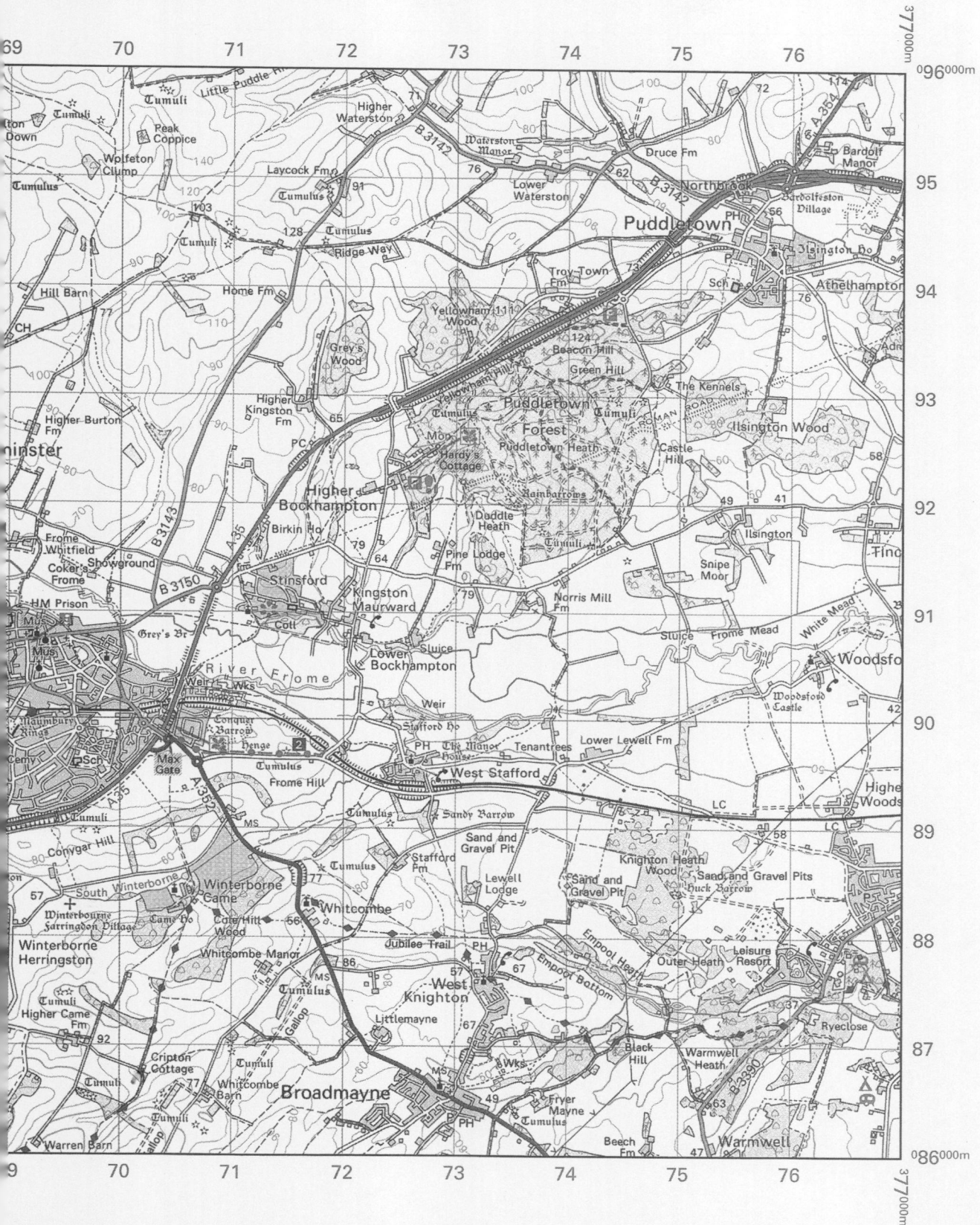

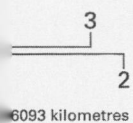

1.

Reference Diagram Q1: The Dorchester Area

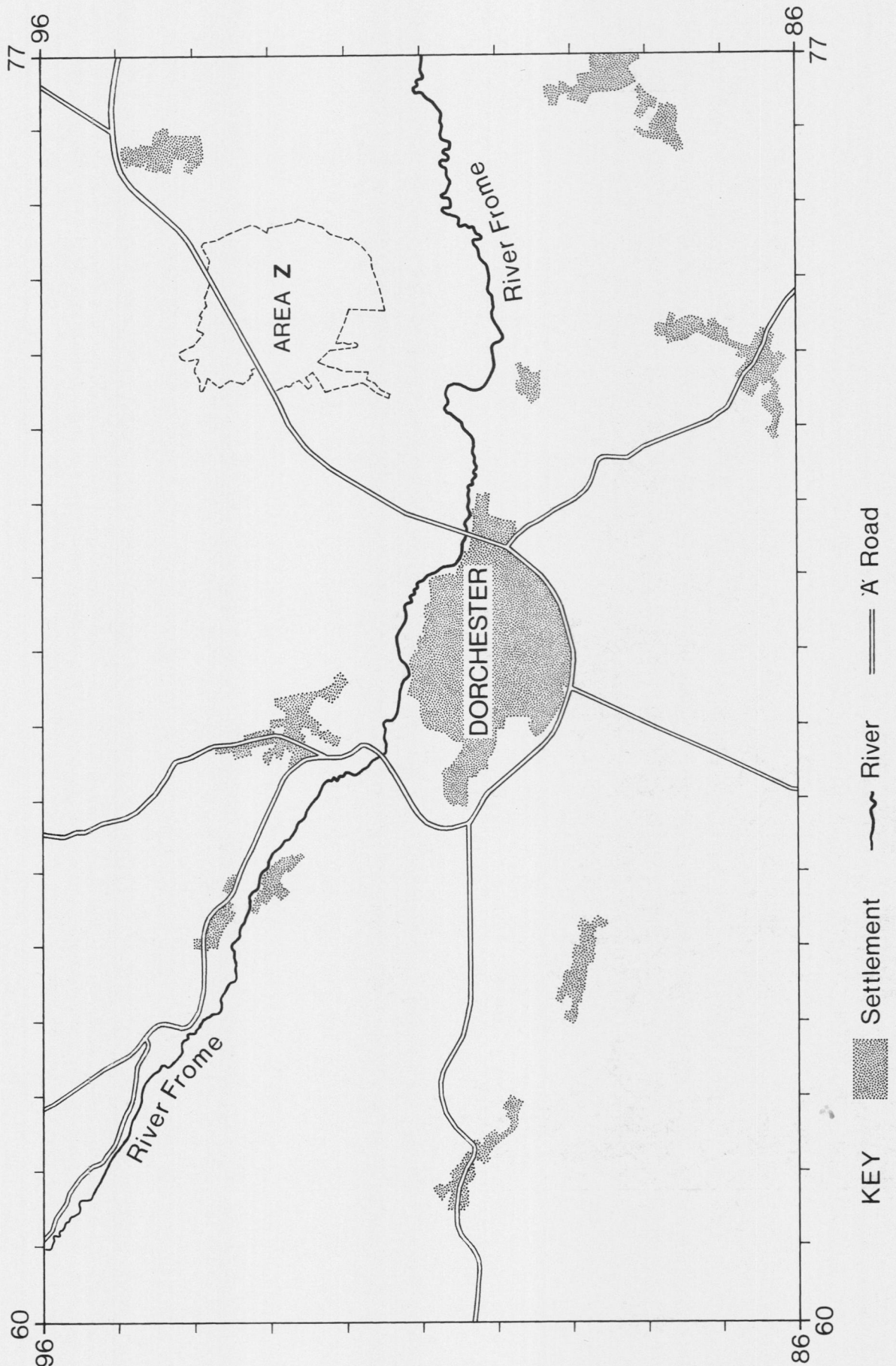

96
77

77
86

River Frome

AREA Z

DORCHESTER

River Frome

KEY

Settlement

River

'A' Road

96
60

60
86

Marks

1. (continued)

Look at the Ordnance Survey Map Extract (No 1557/194) of the Dorchester area and Reference Diagram Q1 on Page two.

(*a*) Complete the table below by matching the physical features to the correct grid references.

Choose from: 6586   7490   6193   6286

Physical Feature	Grid Square
Steep southwest facing slopes	
Flat land	
Broad ridge running East–West	
V-shaped valley	

3

(*b*) Describe the **physical** features of the River Frome **and** its valley between grid references 610959 and 700909.

_____

_____

_____

_____

_____

_____

4

[Turn over

DO NOT
WRITE IN
THIS
MARGIN

KU E

Marks

1. (continued)

(c) Give **two** techniques which could be used to gather information about the physical characteristics of the River Frome.

Give reasons for your choice of techniques.

Technique 1: _____

Reason: _____

_____

Technique 2: _____

Reason: _____

_____

4

(d) It is proposed to develop Area **Z** into a country park (see Reference Diagram Q1).

Using map evidence, give arguments for **and** against this proposal.

For: _____

_____

_____

_____

Against: _____

_____

_____

_____

4

Marks

1. (continued)

(*e*) **Explain** why Dorchester has expanded west into grid square 6790 and not north into grid square 6991.

Refer to both grid squares in your answer.

_____

_____

_____

_____

_____

_____

3

(*f*) Which is the more likely function of Dorchester?

Tick (✓) your choice.

Tourist resort ☐   Market town ☐

Give map evidence to support your choice.

_____

_____

_____

_____

_____

3

(*g*) What are the **disadvantages** of the location of Maiden Castle Farm (grid square 6789)?

_____

_____

_____

_____

_____

3

Marks

2.   **Reference Diagram Q2: A Glaciated Lowland**

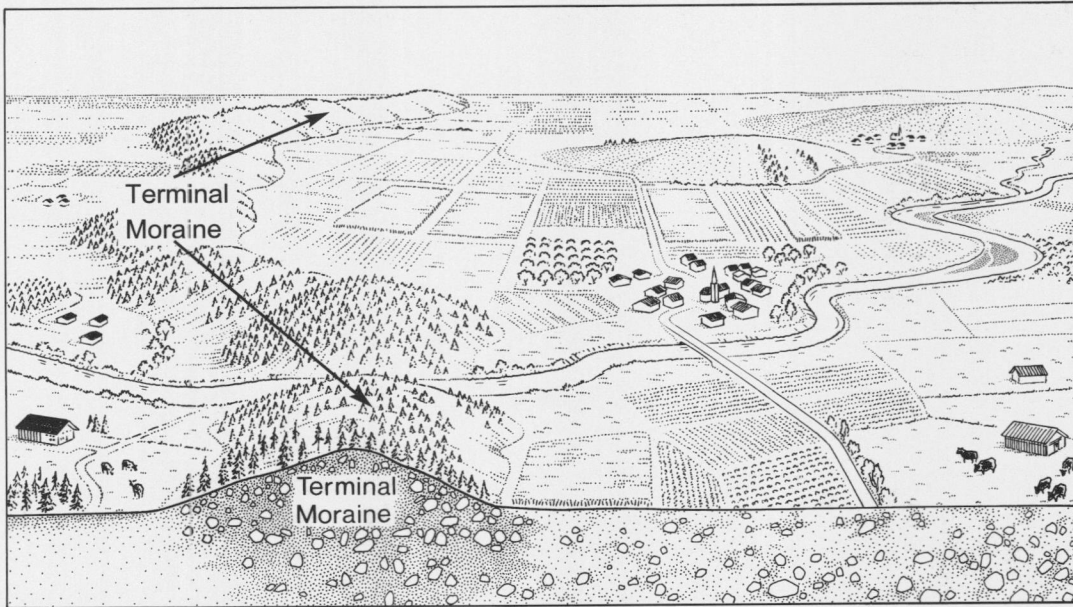

Look at Reference Diagram Q2.

Explain how a terminal moraine is formed.

You may wish to use diagram(s) to illustrate your answer.

_____

_____

_____

_____

(*Space for diagrams*)

3

Marks

3. **Reference Diagram Q3: Air Masses affecting the British Isles**

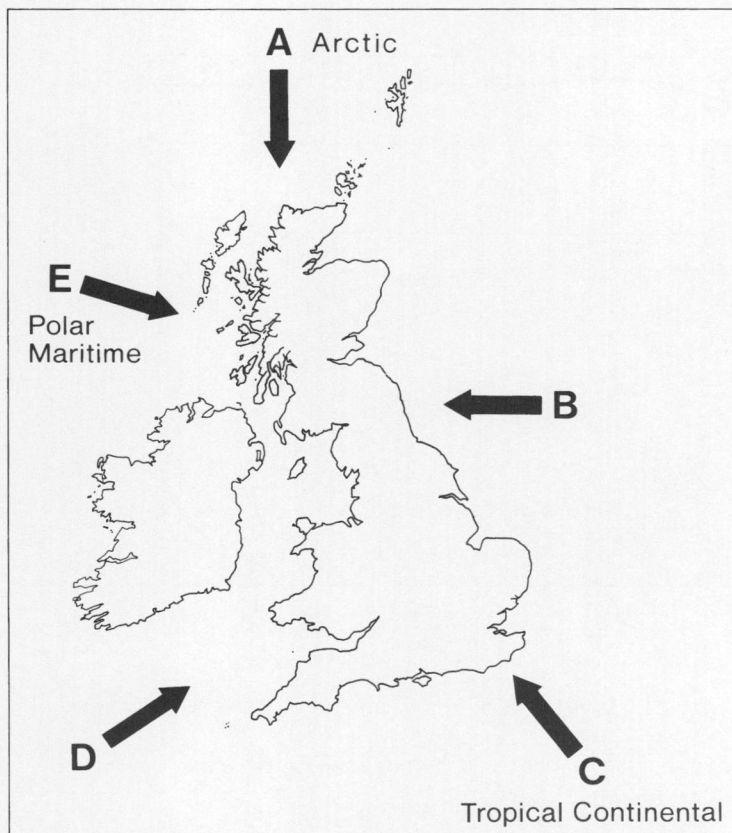

Look at Reference Diagram Q3.

(*a*)   Name air masses B and D.

B _____

D _____

(*b*)   **Describe** the benefits **and** problems of a long spell of weather caused by air mass C in summer.

_____

_____

_____

_____

_____

2

4

[Turn over

Marks KU ES

4.   Reference Diagram Q4A: A Tropical Rainforest Climate Graph

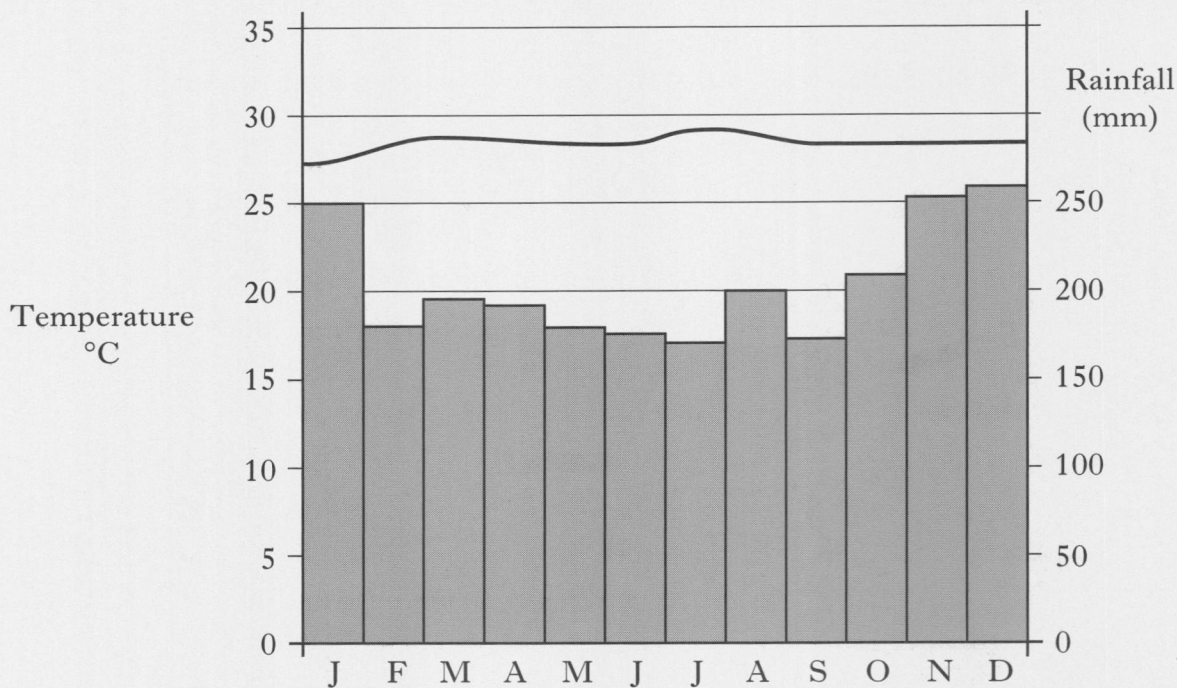

(*a*)   Look at Reference Diagram Q4A.

Describe, **in detail**, the climate of a tropical rainforest.

_____

_____

_____

_____

_____

_____

3

4. (continued)

Reference Diagram Q4B: Developments in the Rainforests of Brazil

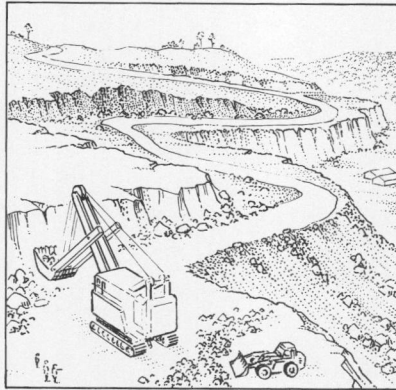

(b)   Look at Reference Diagram Q4B.

"Developments in rainforests have brought many benefits to local people."

Do you agree with the above statement?

Explain your answer.

_____

_____

_____

_____

_____

_____

_____

4

[Turn over

5.   **Reference Diagram Q5A: Brockan Farm in 1977**

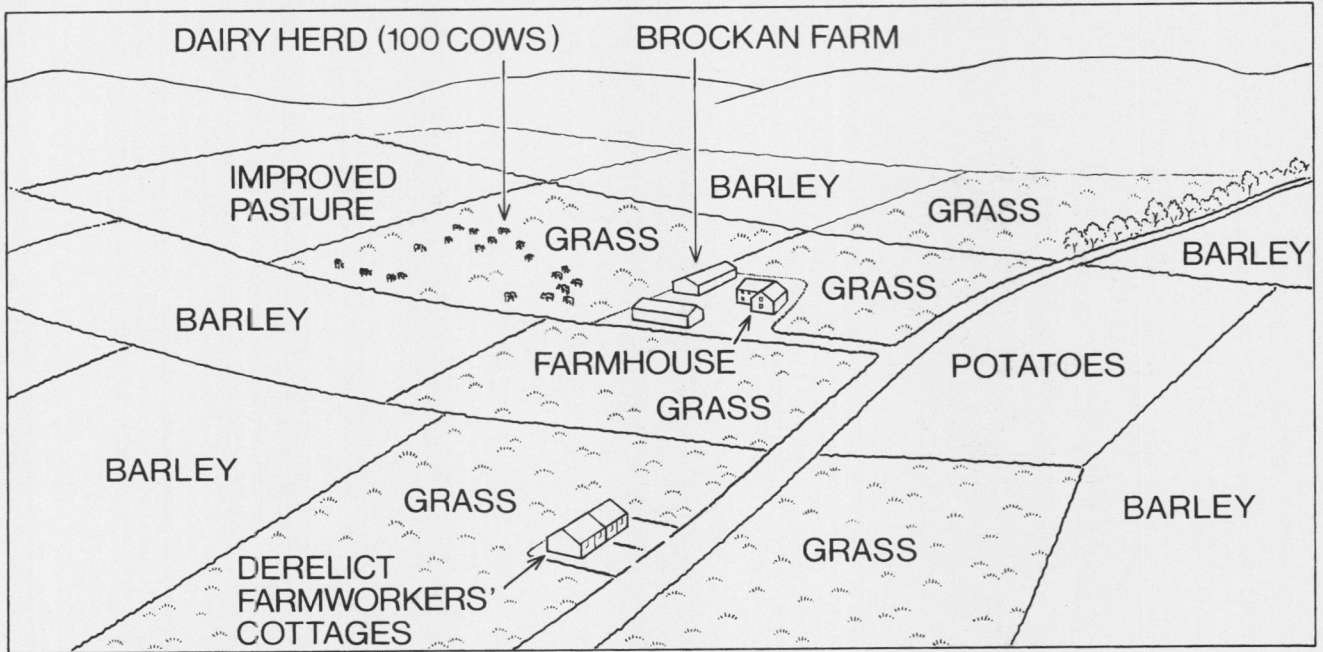

DAIRY HERD (100 COWS)   BROCKAN FARM

IMPROVED PASTURE   BARLEY   GRASS

GRASS   BARLEY

BARLEY   GRASS

FARMHOUSE   POTATOES   GRASS

BARLEY   GRASS   BARLEY

DERELICT FARMWORKERS' COTTAGES   GRASS

Reference Diagram Q5B: Brockan Farm in 2007

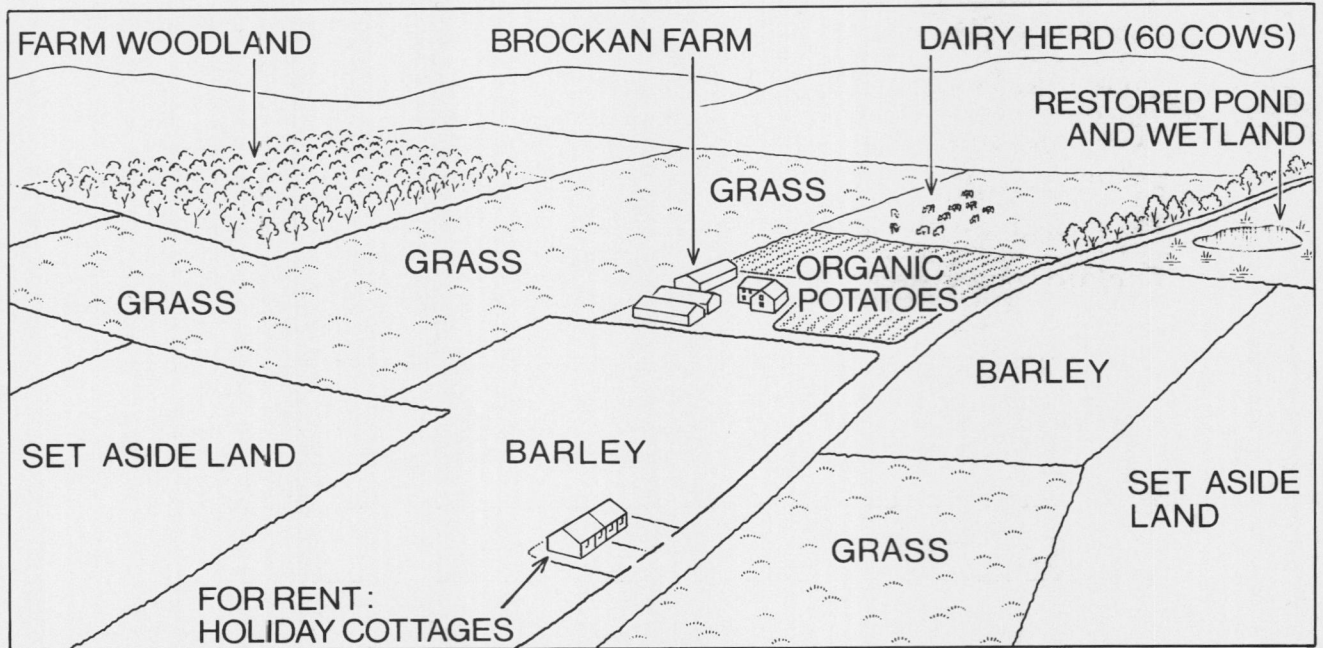

FARM WOODLAND   BROCKAN FARM   DAIRY HERD (60 COWS)

RESTORED POND AND WETLAND

GRASS

GRASS   ORGANIC POTATOES

GRASS   BARLEY

SET ASIDE LAND   BARLEY   SET ASIDE LAND

FOR RENT: HOLIDAY COTTAGES   GRASS

Marks

5. (continued)

(a) Study Reference Diagrams Q5A and Q5B. Suggest reasons for the changes shown on Brockan Farm between 1977 and 2007.

_____

_____

_____

_____

_____

_____

4

Reference Diagram Q5C: Land Use on Brockan Farm, 2007

Land Use	% of Total Land
Grass	45%
Barley	25%
Woodland	10%
Set aside	10%
Organic potatoes	5%
Restored wetland	5%

(b) Study Reference Diagram Q5C.

Complete the divided bar graph below.

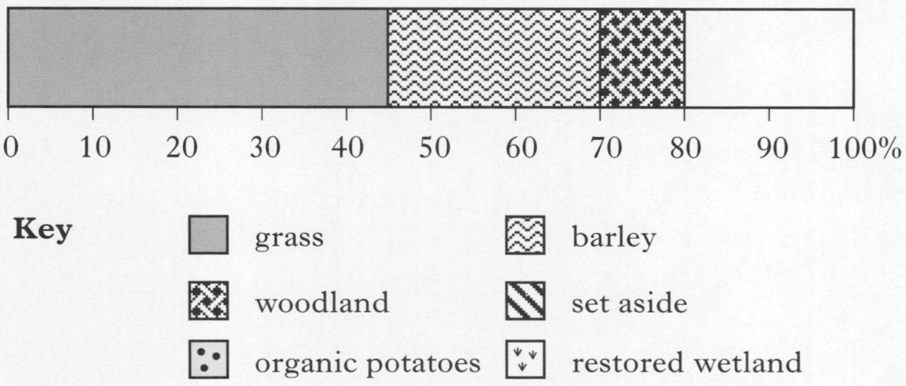

0 10 20 30 40 50 60 70 80 90 100%

Key   grass   barley

             woodland   set aside

             organic potatoes   restored wetland

3

[Turn over

6. **Reference Diagram Q6: Location of a Cement Works**

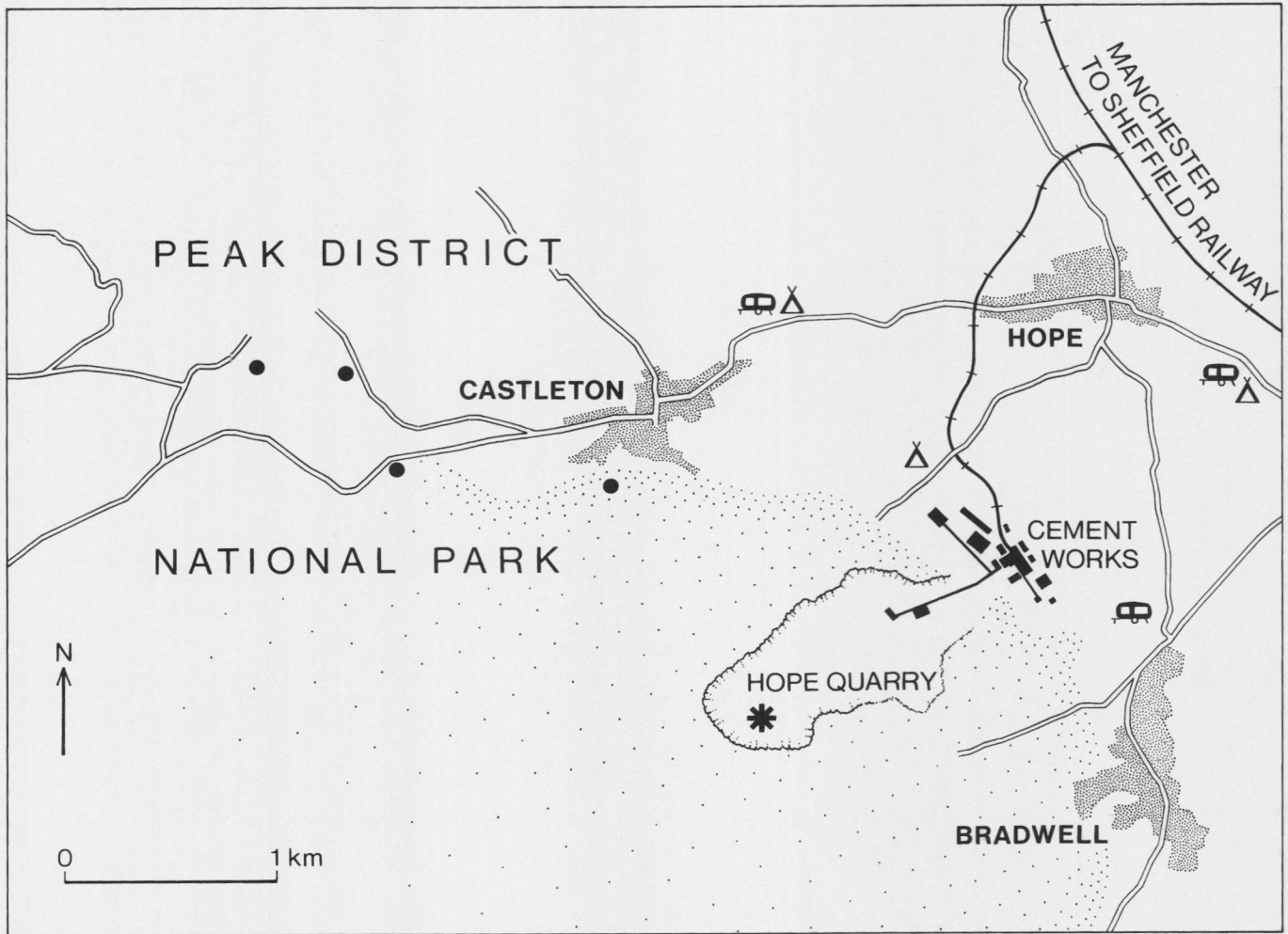

✳ LIMESTONE, THE MAIN RAW MATERIAL IN THE MANUFACTURE OF CEMENT

KEY
● LIMESTONE CAVES (OPEN TO PUBLIC)
🚐 CARAVAN SITE
△ CAMPSITE
▦ BUILT UP AREA

┼─┼─ RAILWAY
══ ROAD
▦ LIMESTONE HILLS

DO NOT WRITE IN THIS MARGIN

KU	ES

Marks

6. (continued)

Study Reference Diagram Q6.

Do you think this is a good location for a cement works?

Give reasons for your answer.

4

[Turn over

Marks

7. **Reference Diagram Q7: Population Data—India**

Year	Population in Millions
1945	336
1955	395
1965	482
1975	600
1985	749
1995	934
2005	1095

Look at Reference Diagram Q7.

(*a*)

"In an Economically Less Developed Country such as India, the population figures taken from census records are likely to be unreliable."

UN spokesperson

Do you agree with the above statement?

Give reasons for your answer.

_____

_____

_____

_____

_____

_____

_____

3

DO NOT WRITE IN THIS MARGIN

KU | ES

Marks

7. (continued)

(*b*) What other techniques could be used to show the information in Reference Diagram Q7?

Explain your choice(s).

4

[Turn over

Marks

8. **Reference Diagram Q8: Migration from Haiti to USA, 1990–95**

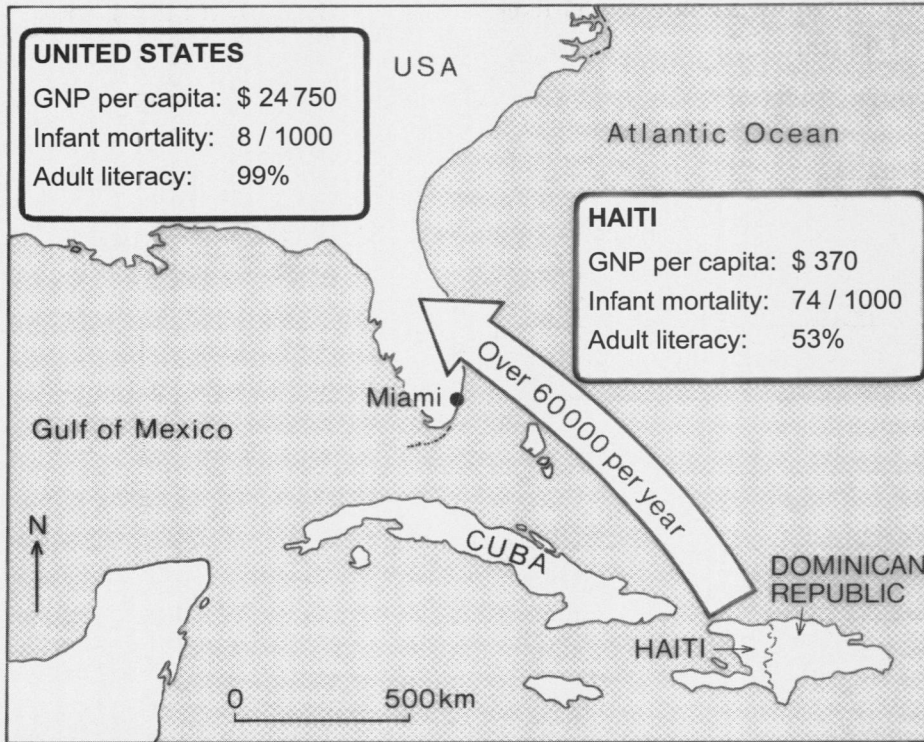

Look at Reference Diagram Q8.

Referring to the data shown, **explain** why so many people migrated from Haiti to USA between 1990 and 1995.

4

9.   **Reference Diagram Q9: Selected World Oil Consumption, 2005**

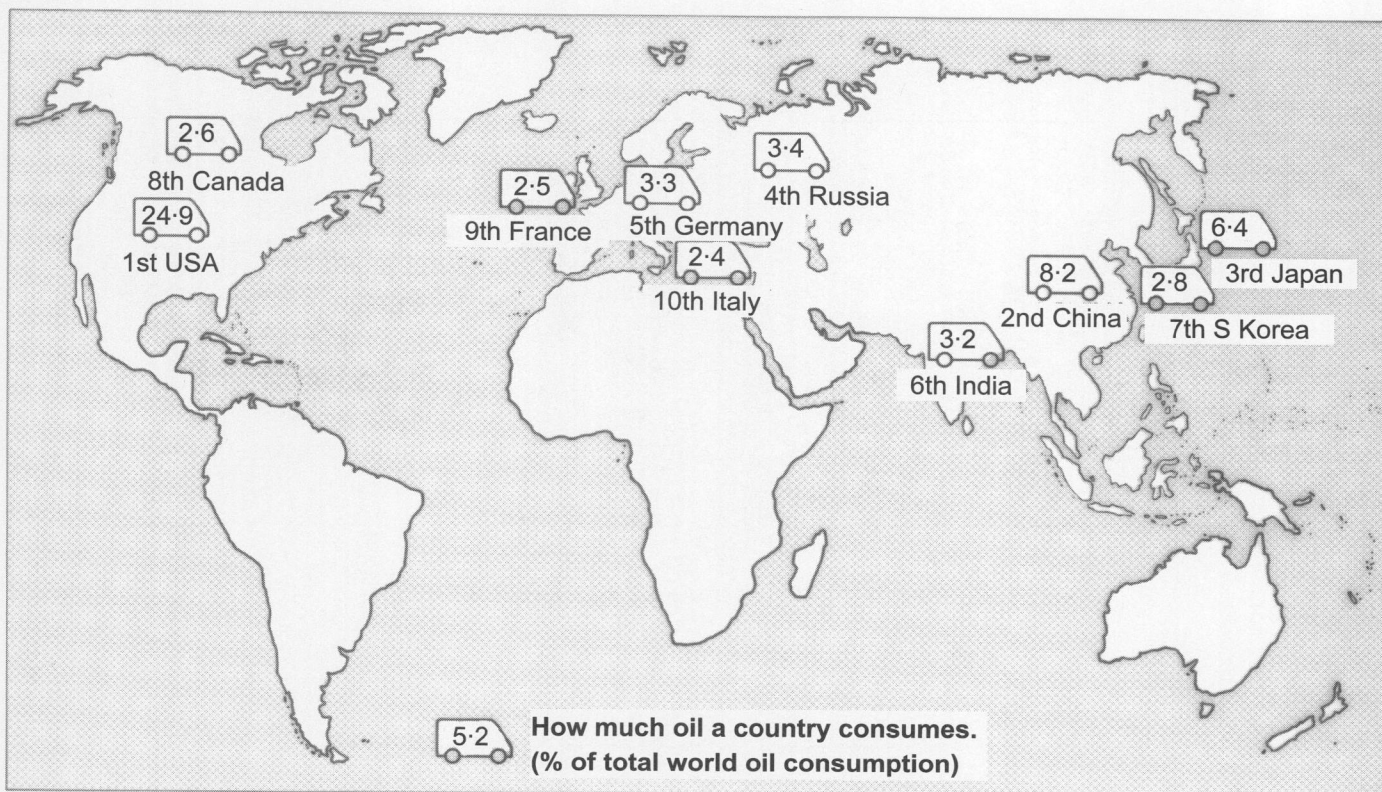

How much oil a country consumes.
(% of total world oil consumption)

Study Reference Diagram Q9.

Explain why certain areas such as the USA, Europe and Japan use such a great amount of the world's oil.

Marks

4

[Turn over

10. **Reference Diagram Q10: Effects of Asian Tsunami, December 2004**

Damage in Nam Khem village, Thailand

Devastation at Petang beach resort, Thailand

Farmland destroyed, Sri Lanka

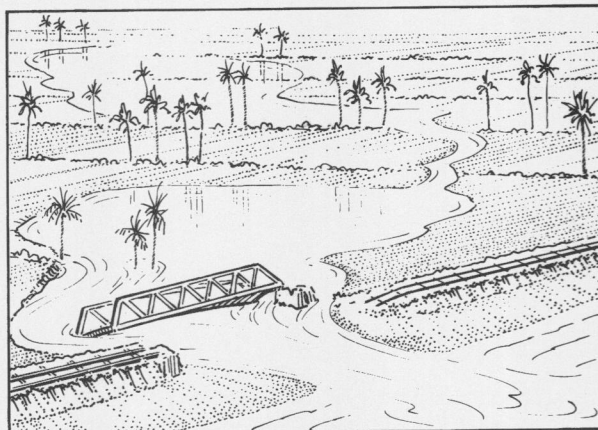

Marks

10. (continued)

Study Reference Diagram Q10.

"Immediate help is essential but we also need long term aid for a full recovery."

Government spokesperson

For this type of natural disaster, **describe** what could be done to help these areas **in the longer term**.

_____

_____

_____

_____

_____

_____

4

[END OF QUESTION PAPER]

[BLANK PAGE]

[BLANK PAGE]

C

1260/405

NATIONAL QUALIFICATIONS 2007	TUESDAY, 8 MAY 1.00 PM – 3.00 PM	GEOGRAPHY STANDARD GRADE Credit Level

All questions should be attempted.

Candidates should read the questions carefully. Answers should be clearly expressed and relevant.

Credit will always be given for appropriate sketch-maps and diagrams.

Write legibly and neatly, and leave a space of about one centimetre between the lines.

All maps and diagrams in this paper have been printed in black only: no other colours have been used.

SCOTTISH QUALIFICATIONS AUTHORITY

1:50 000 Scale
Landranger Series

1 kilometre = 0·6214 mile

Extract No 1558/54

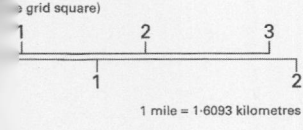

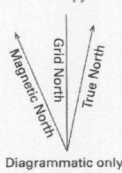

DUNDEE

TAY

BROUGHTY FERRY

Barnhill

TAYPORT

St Michaels

Pile Lighthouse

Morton Links

1 mile = 1·6093 kilometres

Diagrammatic only

Grid North
True North
Magnetic North

1.

Reference Diagram Q1A

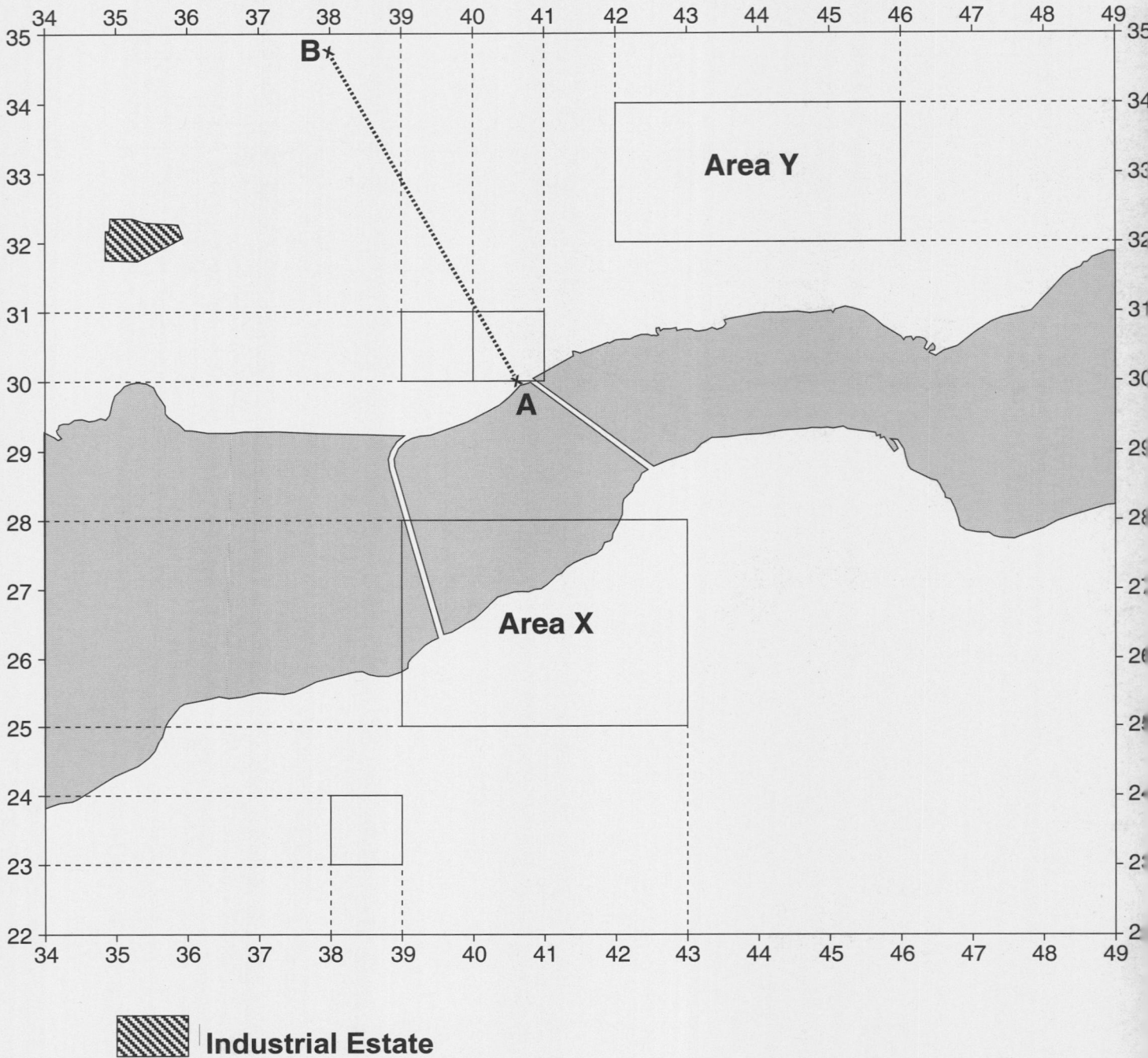

Industrial Estate

Marks

KU	ES

1. (continued)

This question refers to the OS Map Extract (No 1558/54) of the Dundee area.

Reference Diagram Q1B: View SE from Dundee Law 392313

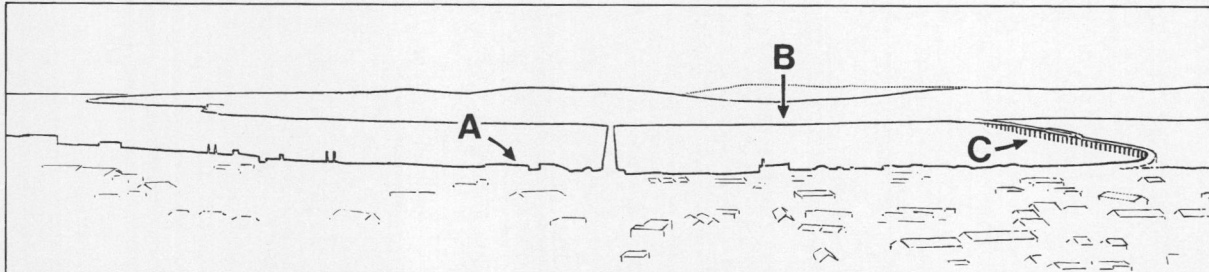

(a) Study Reference Diagram Q1B and the map extract.

Identify the **three** features A, B and C.

Choose from:

Railway Bridge;   Tayport;   Discovery Point;   Road Bridge;

Docks;   Newport on Tay.

3

(b) Find Area X on Reference Diagram Q1A (see *Page two*) and on the OS map extract.

Referring to map evidence, **explain** the way in which the **physical** landscape has affected land use in this area.

4

[Turn over

Mar

KU

1. (continued)

Reference Diagram Q1C: Area Y in the Year 1971

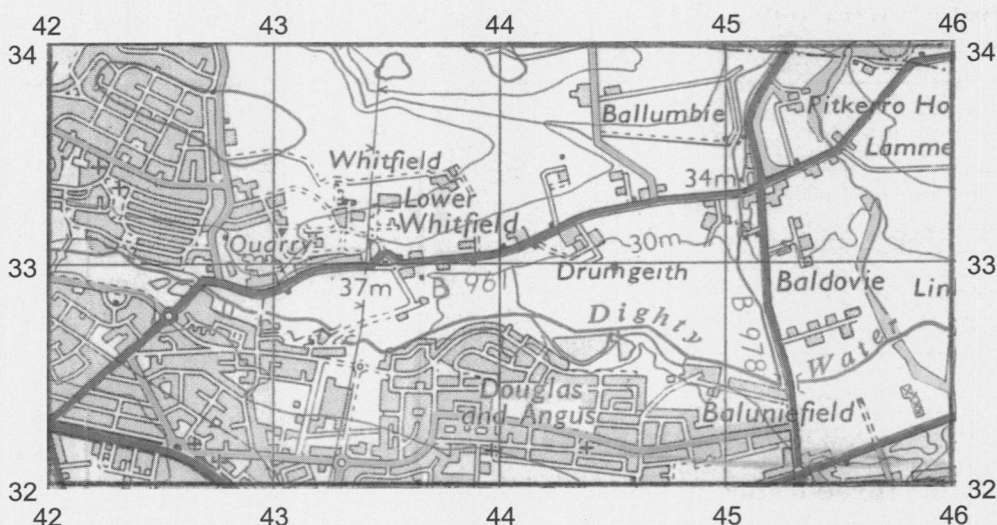

(c) Look at Reference Diagram Q1C.

Find Area Y on Reference Diagram Q1A and on the OS map extract.

Describe how the land use changes in the area since 1971 have **both** benefited **and** created problems for the area and its people.

(d) Mr Dick works in grid square 4030 and lives in a flat in grid square 3930. He is considering moving house to Gauldry 3823.

Do you think he should make this move?

Using map evidence, give reasons for your answer.

(e) Refer to Reference Diagram Q1A.

A group of students intends to gather information about differences in urban land use along transect AB (407300 to 380348).

Describe, in detail, the gathering techniques they might use.

Give reasons for your choice of techniques.

(f) **Explain** the location of the industrial estate at 3532.

4

Marks

KU ES

2.   **Reference Diagram Q2: A Hanging Valley**

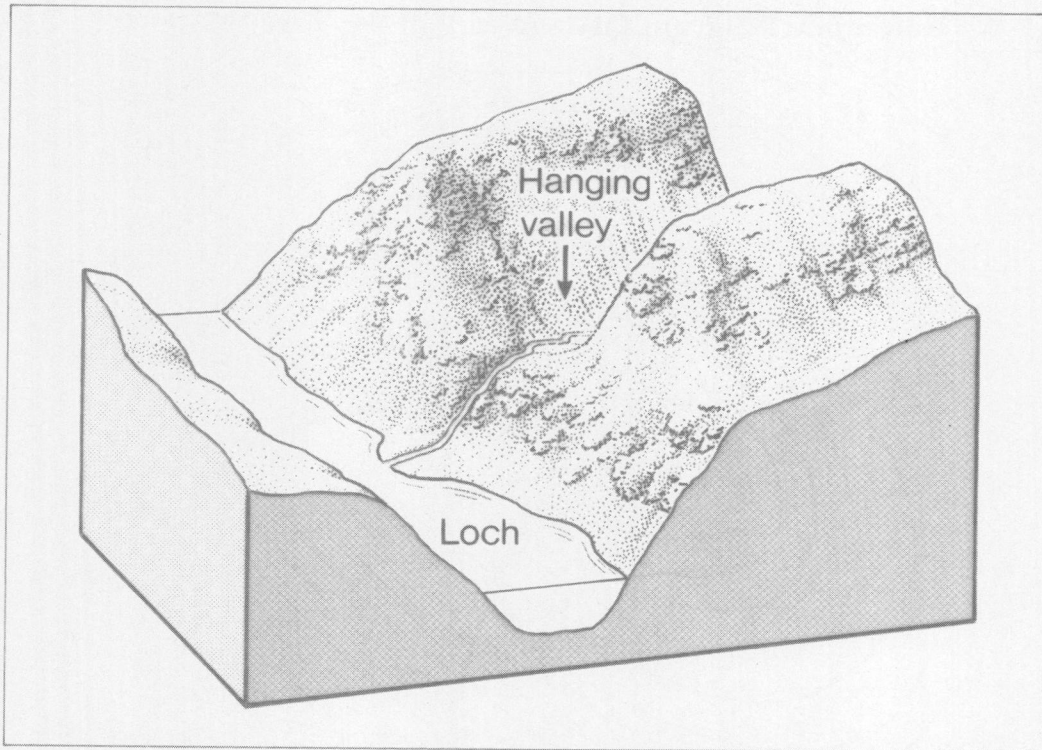

Look at Reference Diagram Q2.

Explain how a hanging valley is formed.

You may use diagram(s) to illustrate your answer.

4

[Turn over

Mark

KU

3.   **Reference Diagram Q3A: Synoptic Chart 12 noon, 18 November 2006**

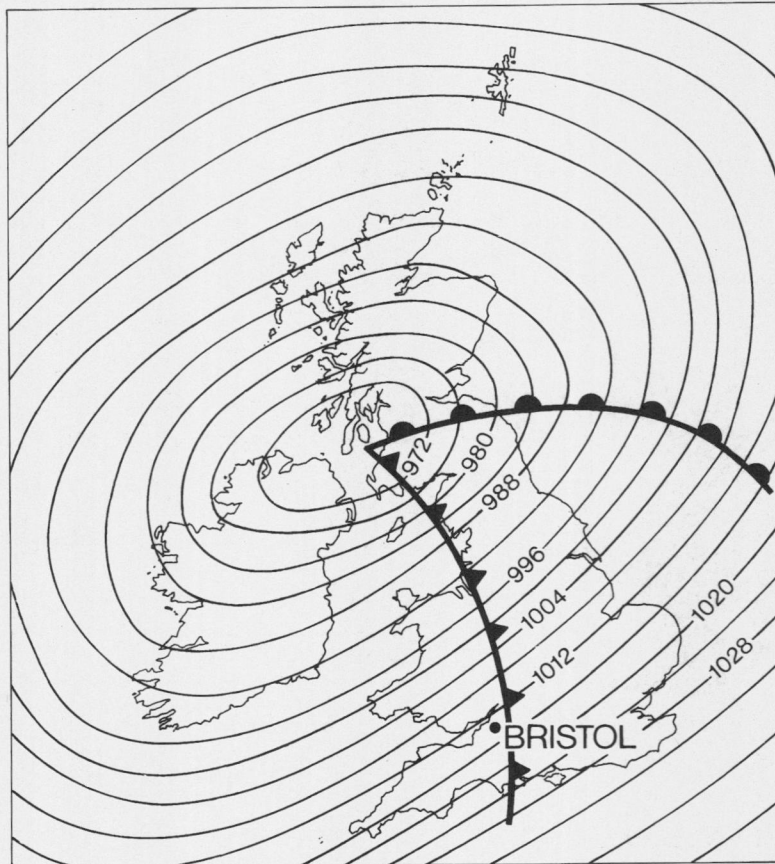

Reference Diagram Q3B: Two Sets of Weather Information

	Set X	Set Y
Temperature	5 °C	12 °C
Wind speed	35 knots	10 knots
Wind direction	SW	E
Precipitation	Heavy rain	Steady rain
Cloud amount	7 oktas	4 oktas
Cloud type	Cumulonimbus	Stratus

Look at Reference Diagrams Q3A and Q3B.

Which set of weather information, X or Y, is correct for Bristol?

Explain your choice in detail.

5

4.   **Reference Diagram Q4: Desertification**

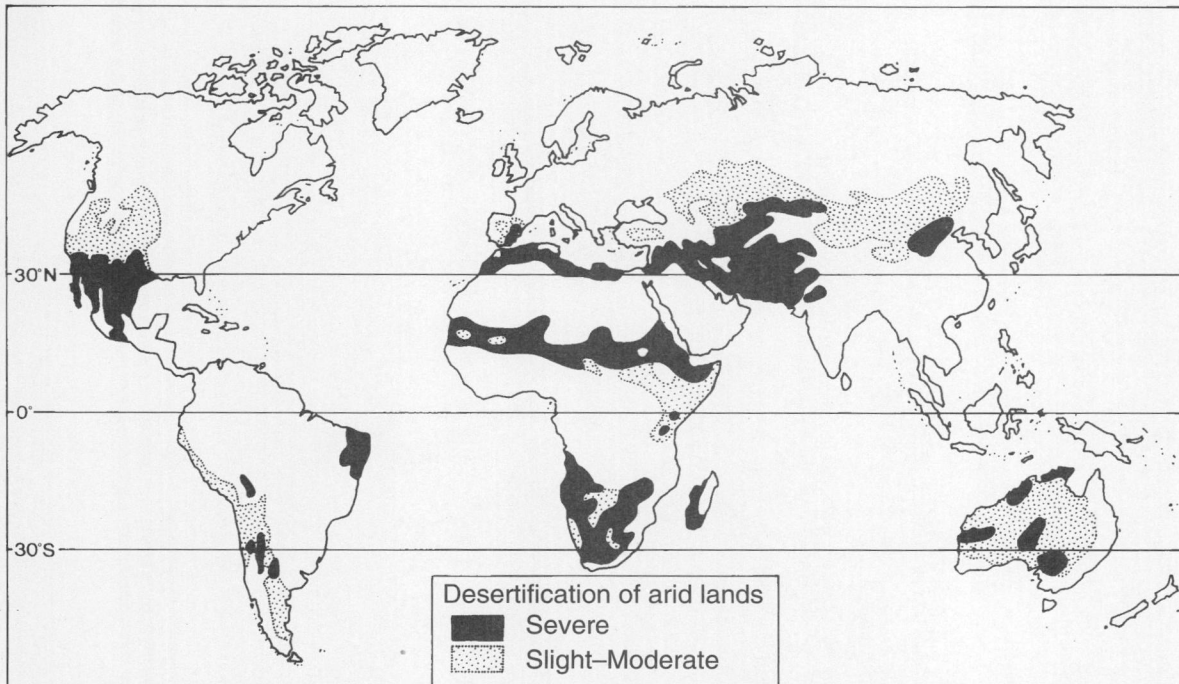

Physical Causes of Desertification	Human Causes of Desertification
Unreliable rainfall	Population increase
Wind	Overgrazing/overcropping
High temperatures	Removing trees for firewood

Look at Reference Diagram Q4.

(a) Desertification is a major problem in many areas of the world.

   Choose **one** physical and **one** human cause and **explain** why each of them is a major reason for desertification.   4

(b) **Describe**, in detail, ways in which desertification can be overcome.   4

[Turn over

Mar
KU

5.   **Reference Diagram Q5: Central Business District of a Large City**

Study Reference Diagram Q5.

In recent years many changes have taken place in the Central Business Districts of British cities.

Give reasons for these changes.

5

Marks
KU | ES

6. **Reference Diagram Q6A: Inverlochlarig Sheep Farm, Perthshire**

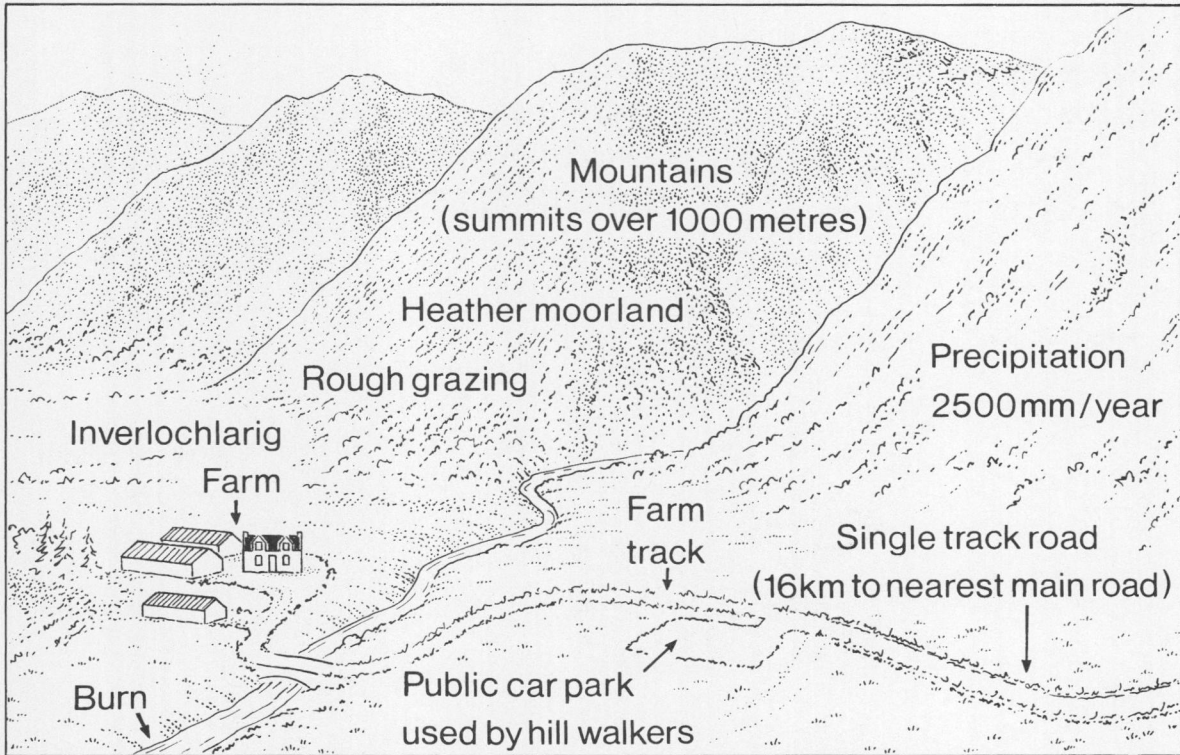

Mountains (summits over 1000 metres)

Heather moorland

Rough grazing

Precipitation 2500 mm/year

Inverlochlarig Farm

Farm track

Single track road (16km to nearest main road)

Public car park used by hill walkers

Burn

Reference Diagram Q6B

"I'm so fed up with ever increasing fuel costs, lower prices from animal sales and dealing with the EU that I wonder if it is worth carrying on."

Local farmer

Study Reference Diagrams Q6A and Q6B.

Human and physical factors both create problems for farmers in this type of environment.

Select **either** human **or** physical factors.

For the factors you have chosen **explain**, in detail, why these cause more problems for the farmer.

6

[Turn over

7.

Reference Diagram Q7A:
The Eden Project—A Visitor Attraction in Cornwall

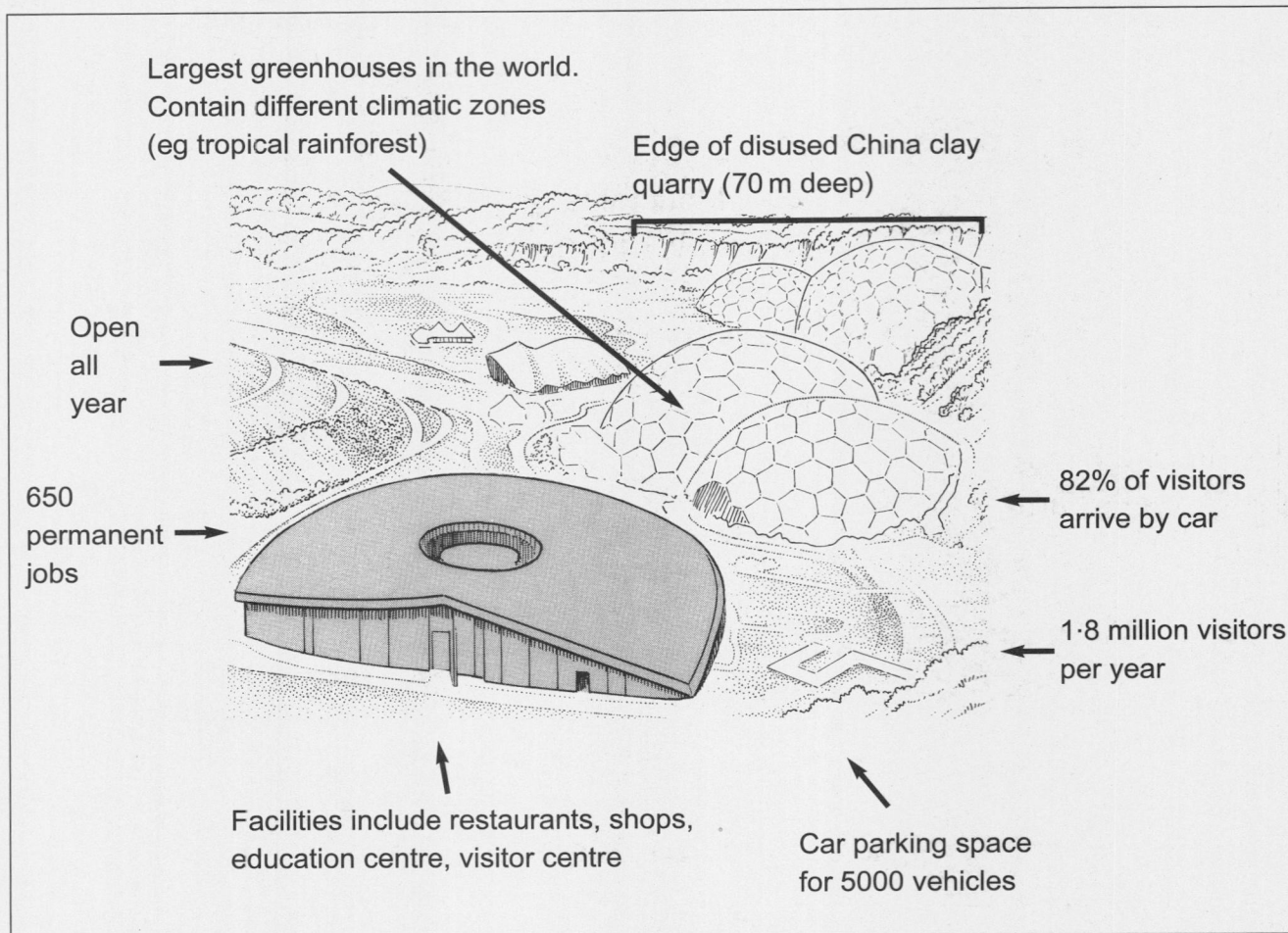

Largest greenhouses in the world.
Contain different climatic zones
(eg tropical rainforest)

Edge of disused China clay
quarry (70 m deep)

Open
all
year

82% of visitors
arrive by car

650
permanent
jobs

1·8 million visitors
per year

Facilities include restaurants, shops,
education centre, visitor centre

Car parking space
for 5000 vehicles

Reference Diagram Q7B: Location of the Eden Project

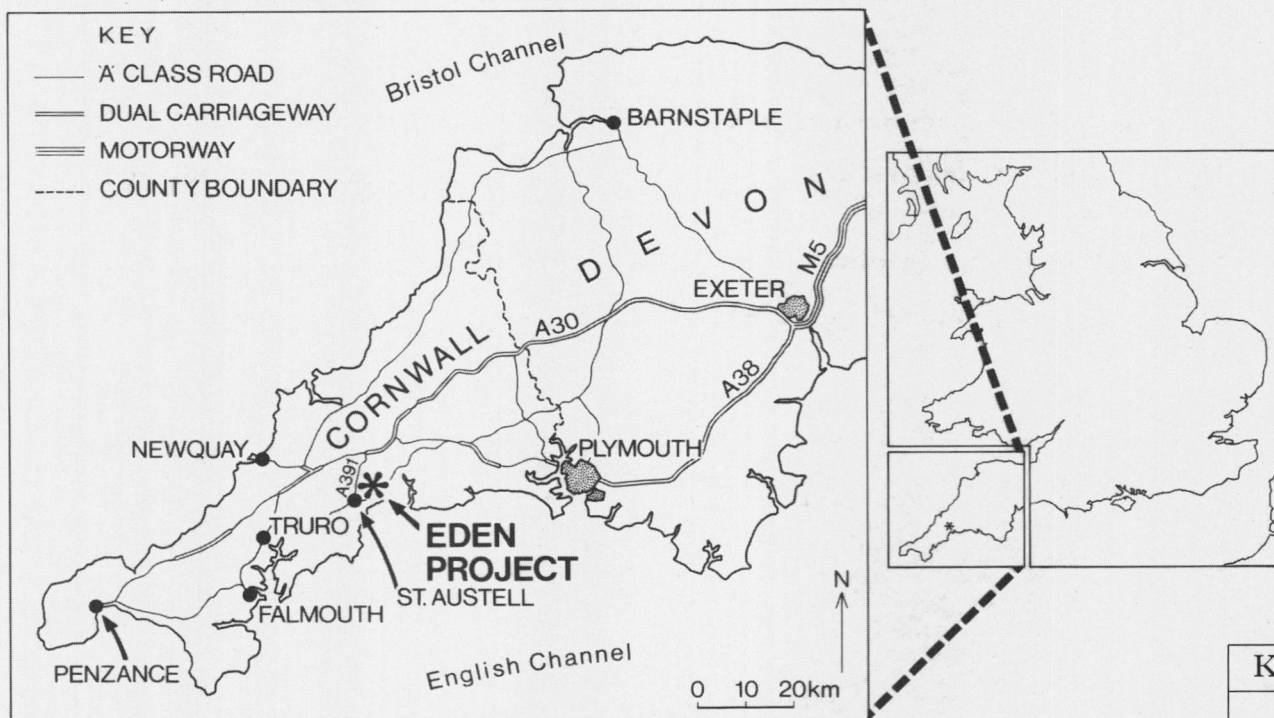

KEY
— 'A' CLASS ROAD
═ DUAL CARRIAGEWAY
≡ MOTORWAY
---- COUNTY BOUNDARY

Bristol Channel

BARNSTAPLE

D E V O N

M5

EXETER

A30

A38

DEVON

CORNWALL

NEWQUAY

A391

PLYMOUTH

TRURO

*

**EDEN
PROJECT**

ST. AUSTELL

FALMOUTH

PENZANCE

N

English Channel

0 10 20km

Study Reference Diagrams Q7A and Q7B.

Explain fully the advantages **and** disadvantages of this new visitor attraction to
St Austell and the surrounding area.

Mar

KU

8. **Reference Diagram Q8A:**
    North America Population Distribution

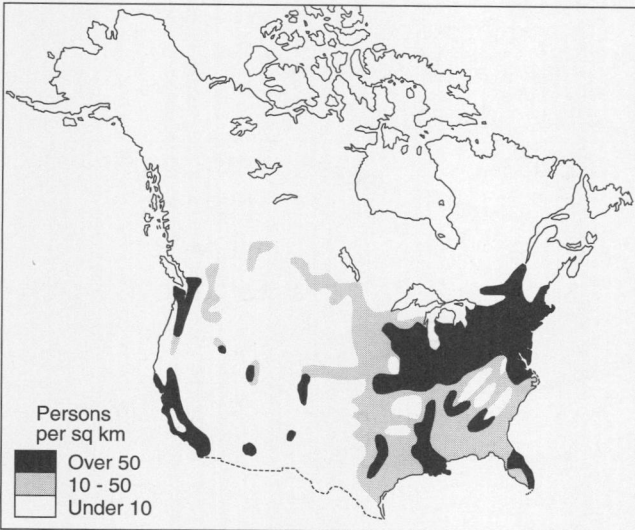

Persons
per sq km
- Over 50
- 10 - 50
- Under 10

Reference Diagram Q8B:
North America Relief

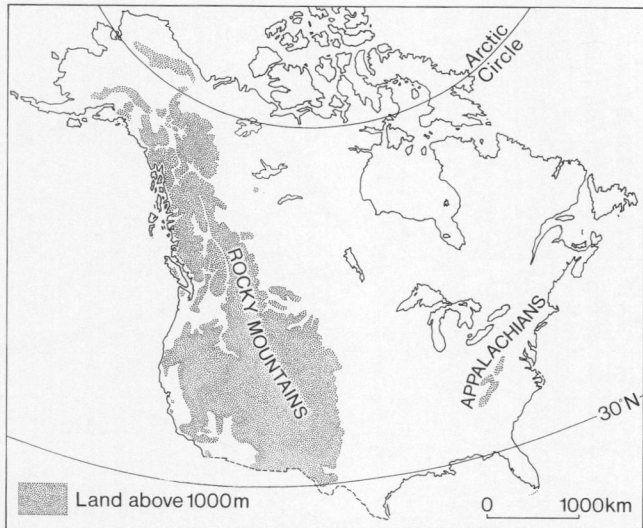

Land above 1000m   0   1000km

Reference Diagram Q8C:
North America Annual Rainfall

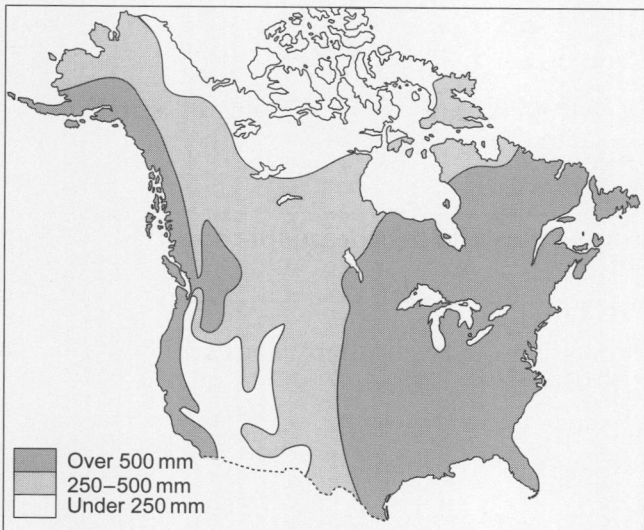

- Over 500 mm
- 250–500 mm
- Under 250 mm

Reference Diagram Q8D:
North America Power and Industry

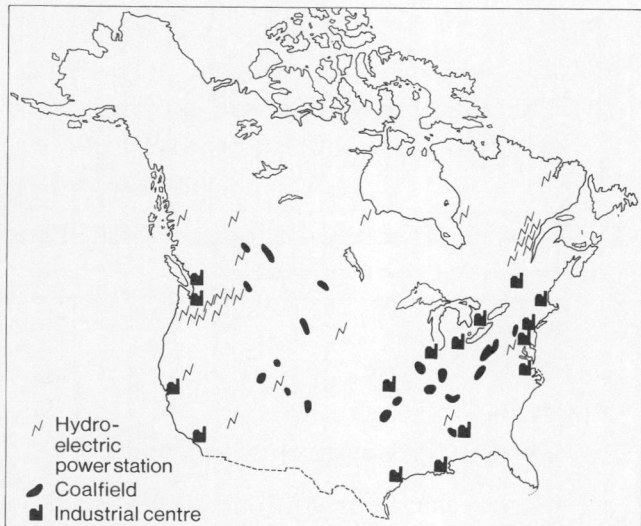

- Hydro-electric power station
- Coalfield
- Industrial centre

| | *Marks* | |
| | KU | ES |

Using information given in Reference Diagrams Q8A, B, C and D, **explain** the distribution of population in North America.

6

[Turn over

9.   **Reference Diagram Q9A: Tourism in the Gambia (West Africa)**

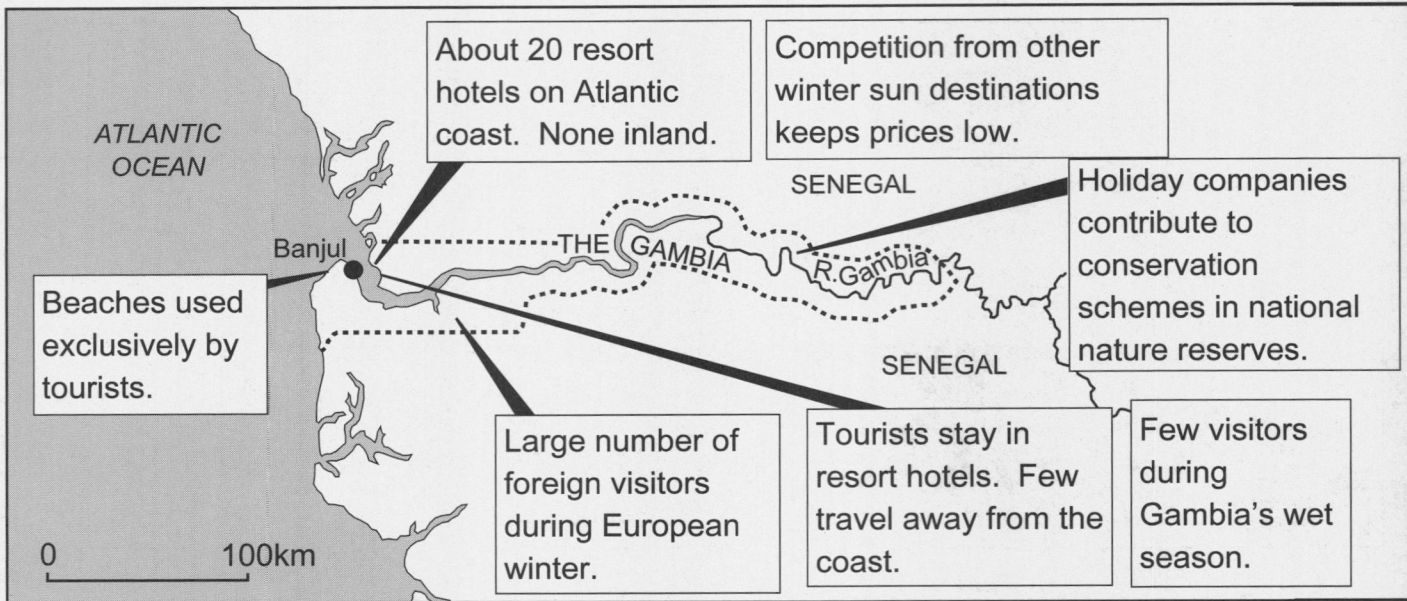

ATLANTIC
OCEAN

About 20 resort hotels on Atlantic coast. None inland.

Competition from other winter sun destinations keeps prices low.

SENEGAL

Holiday companies contribute to conservation schemes in national nature reserves.

Banjul   THE GAMBIA   R.Gambia

Beaches used exclusively by tourists.

SENEGAL

Large number of foreign visitors during European winter.

Tourists stay in resort hotels. Few travel away from the coast.

Few visitors during Gambia's wet season.

0   100km

Mar

KU

(a) The Gambia which is a country in West Africa is known as a "winter sun" destination for tourists from Europe.

"The growth of the tourist industry has brought huge benefits for the people and the environment of the Gambia."

Package holiday company spokesperson

Do you agree with this statement?

Give reasons for your answer.

9. (continued)

Reference Diagram Q9B: Tourism Facts for The Gambia

- Population   =   1 400 000

- Labour Force   =   400 000

- Tourism employs 10 000 Gambians

- Tourism is the biggest foreign exchange earner in The Gambia

- GDP   from Agriculture   = 30%   from Tourism   = 25%

  from Fisheries   = 30%   from Others   = 15%

- Most visitors come from Europe:

  from UK   = 60%   from Sweden = 7%

  from Netherlands = 12%   from Others   = 21%

(b) Study Reference Diagram Q9B.

Which processing techniques would be most effective to show the different information about tourism?

Explain your choice of techniques.

4

[Turn over for Question 10 on *Page fourteen*

10.   **Reference Diagram Q10: Measures of Development**

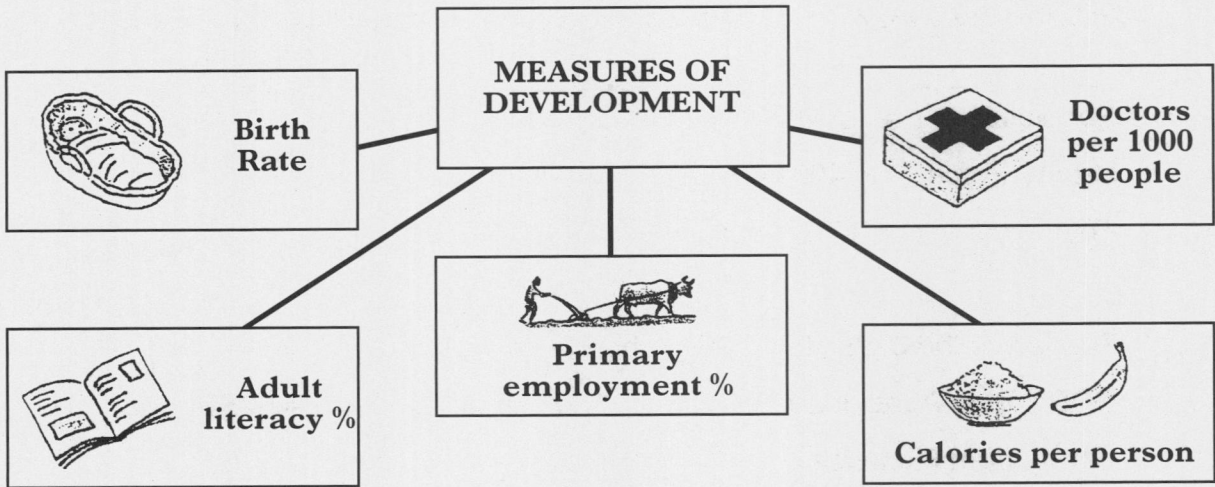

Study Reference Diagram Q10.

Choose **two** of the measures shown and **explain** why they are good indicators of the differences between ELDCs (economically less developed countries) and EMDCs (economically more developed countries).

6

[END OF QUESTION PAPER]

[BLANK PAGE]

FOR OFFICIAL USE

G

KU	ES

Total Marks

1260/403

NATIONAL
QUALIFICATIONS
2008

FRIDAY, 9 MAY
10.25 AM–11.50 AM

GEOGRAPHY
STANDARD GRADE
General Level

Fill in these boxes and read what is printed below.

Full name of centre

Town

Forename(s)

Surname

Date of birth
Day Month Year

Scottish candidate number

Number of seat

1 Read the whole of each question carefully before you answer it.

2 Write in the spaces provided.

3 Where boxes like this ☐ are provided, put a tick ✓ in the box beside the answer you think is correct.

4 Try all the questions.

5 Do not give up the first time you get stuck: you may be able to answer later questions.

6 Extra paper may be obtained from the invigilator, if required.

7 Before leaving the examination room you must give this book to the invigilator. If you do not, you may lose all the marks for this paper.

SQA

Extract No 1655/36

1:50 000 Scale
Landranger Series

Four colours should appear above; if not then please return to the invigilator.
Four colours should appear above; if not then please return to the invigilator.

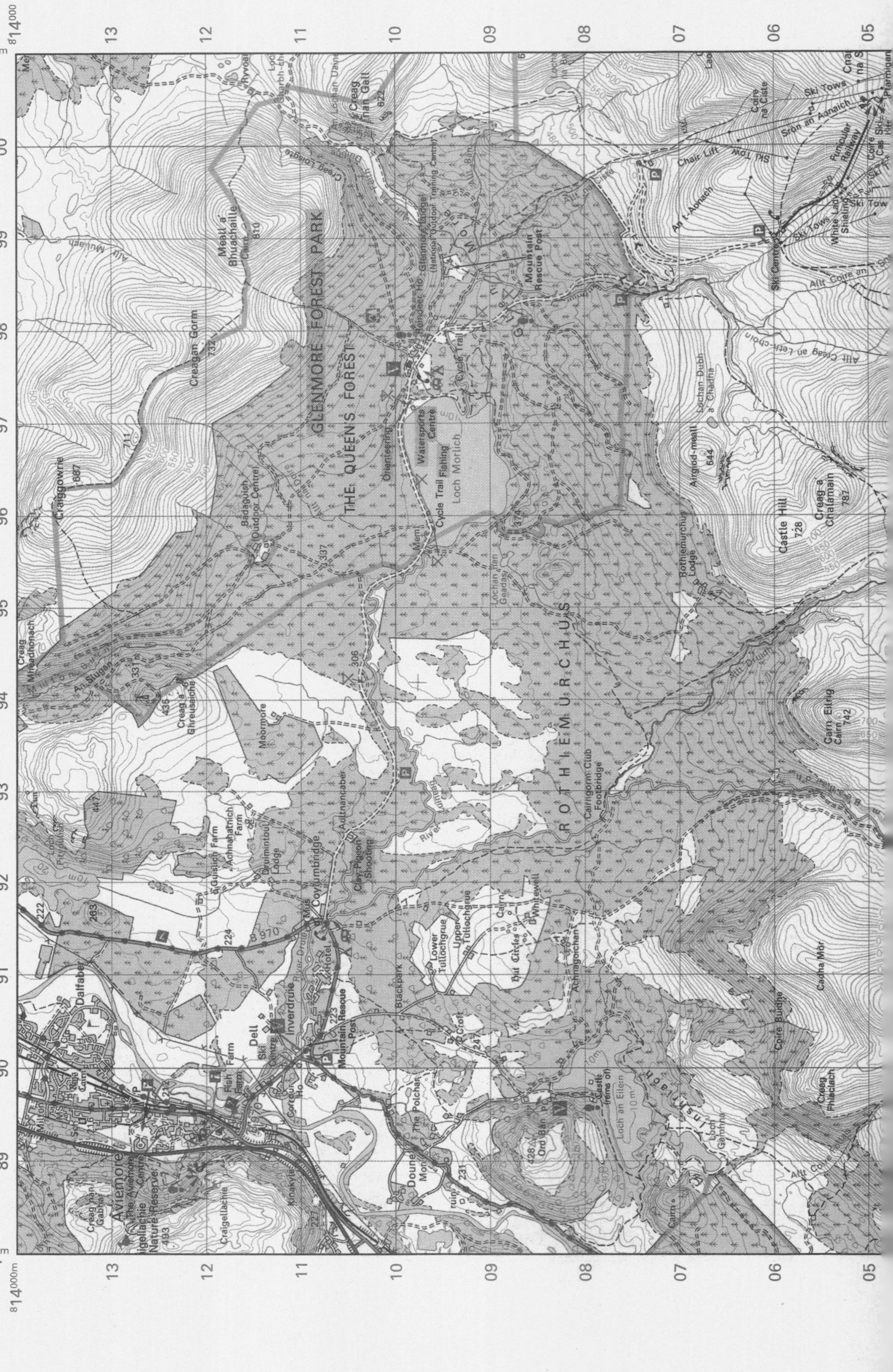

Scale 1: 50 000

2 centimetres to 1 kilometre (one grid square)

3   2   1   0   Kilometres   1   2   3 kilometres

2   1   0   Miles   1   2

1 kilometre = 0·6214 mile

1 mile = 1·6093 kilometres

True North
Grid North
Magnetic North

Diagrammatic only

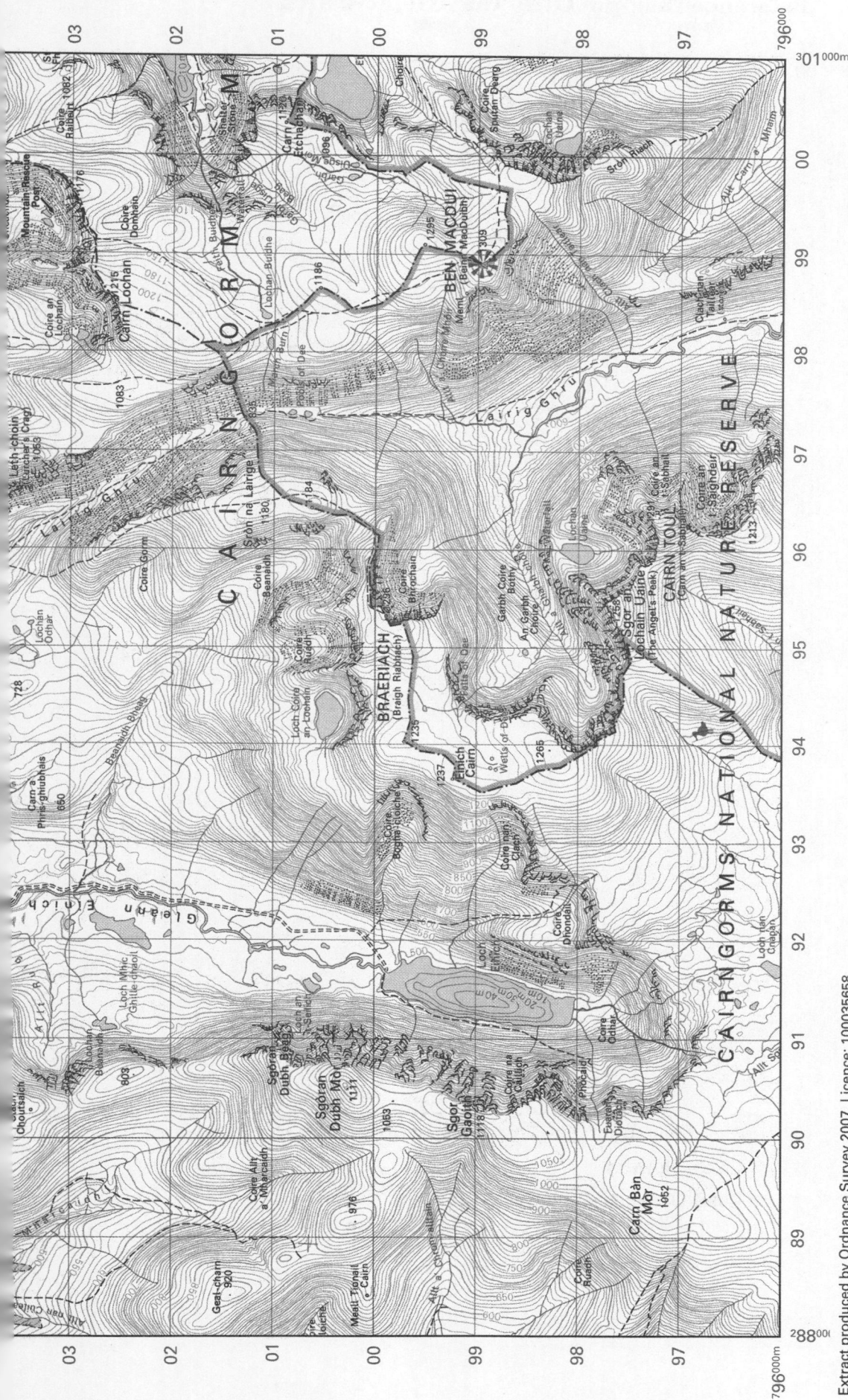

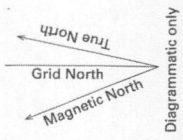

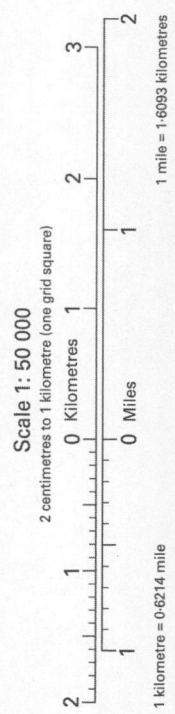

CAIRNGORMS NATIONAL NATURE RESERVE

BEN MACDUI (Beinn MacDuibh)

BRAERIACH (Braigh Riabhach)

CAIRN TOUL (Carn an t-Sabhail)

1.

Reference Diagram Q1A: The Aviemore Area

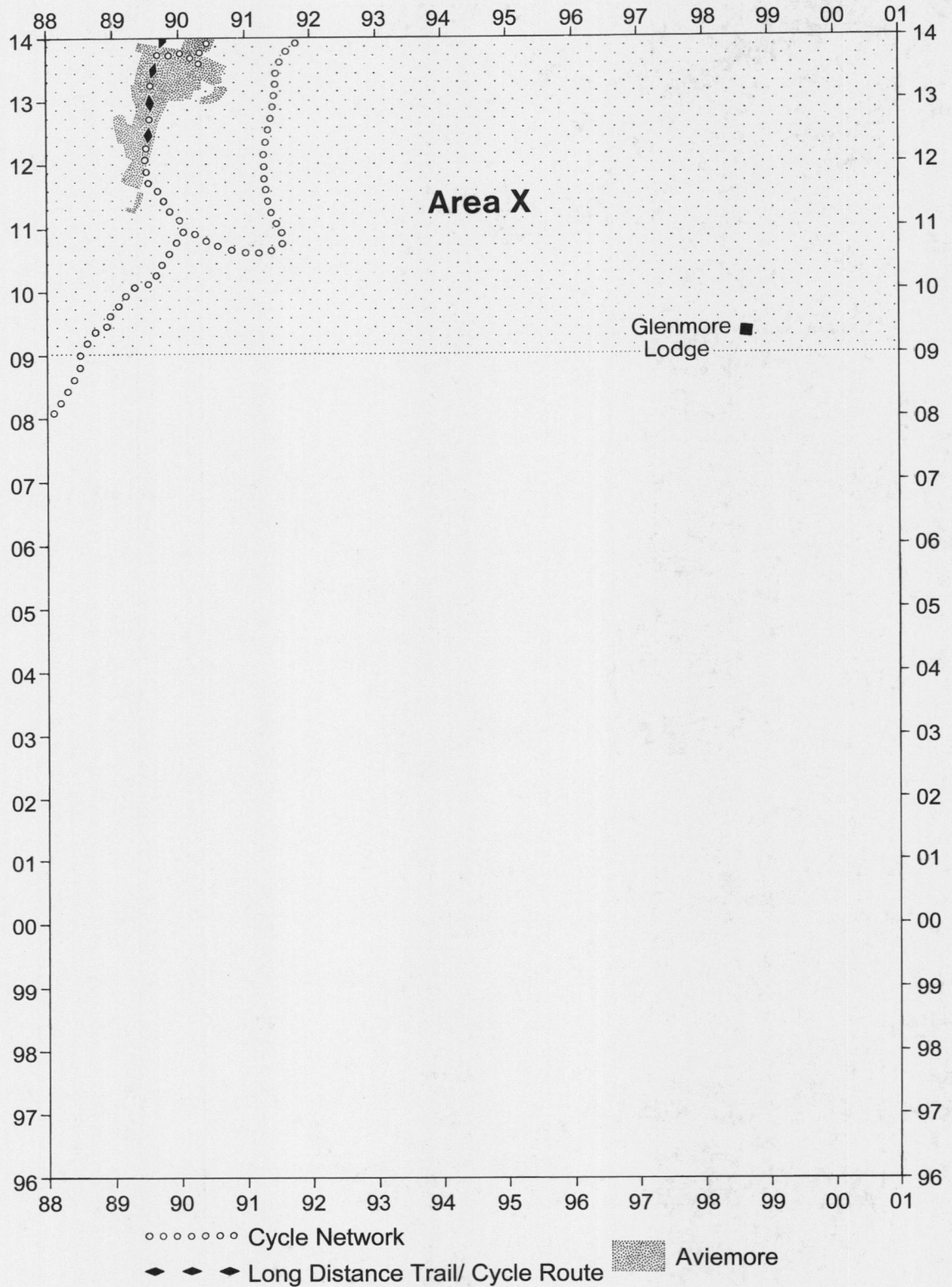

Area X

Glenmore ■
Lodge

∘∘∘∘∘∘∘ Cycle Network

◆ ◆ ◆ Long Distance Trail/ Cycle Route

Aviemore

Marks

1. (continued)

Look at the Ordnance Survey Map Extract (No 1655/36) of the Aviemore area and Reference Diagram Q1A on Page two.

(a) Using the map extract, match the glaciated features A, B, C and D to the correct name/grid reference in the table below.

A Ribbon Lake   B Corrie   C U shaped Valley   D Pyramidal Peak

Grid Reference/Name of Feature	Letter
9597 Angel's Peak	
9798 Lairig Ghru	
9400 Loch Coire an Lochain	
9198 Loch Einich	

3

(b) **Explain** how a pyramidal peak was formed.

You may use diagram(s) to illustrate your answer.

_____

_____

_____

_____

_____

_____

_____

_____

_____

_____

3

Marks

1. (continued)

(c)   Find Area X on the map extract and on Reference Diagram Q1A.

In what ways has the physical landscape in Area X both encouraged **and** limited settlement growth?

_____

_____

_____

_____

_____   **4**

(d)   Glenmore Lodge (9809) is a National Outdoor Training Centre*.

(*This type of centre provides instruction in outdoor activities.)

How good is this location for the centre?

Give reasons for your answer.

_____

_____

_____

_____

_____   **4**

Marks

1. (continued)

Reference Diagram Q1B: Selected Aims of National Parks

- *Preserve beauty of countryside*
- *Conserve local wildlife*
- *Provide good access and facilities for public enjoyment*
- *Maintain farming*

(*e*)   This area is part of the Cairngorm National Park.

Explain how land uses shown on the OS map **and** outdoor activities in this area might be in conflict with the aims shown in Reference Diagram Q1B.

_____

_____

_____

_____

_____

4

(*f*)   There is a Long Distance Trail and Cycle Network shown on both the map extract and Reference Diagram Q1A.

What techniques could be used to gather information on the impact of walkers and cyclists on the local area?

Give reasons for your choices.

_____

_____

_____

_____

_____

_____

4

[Turn over

Marks

2.   **Reference Diagram Q2: A Lowland River Landscape**

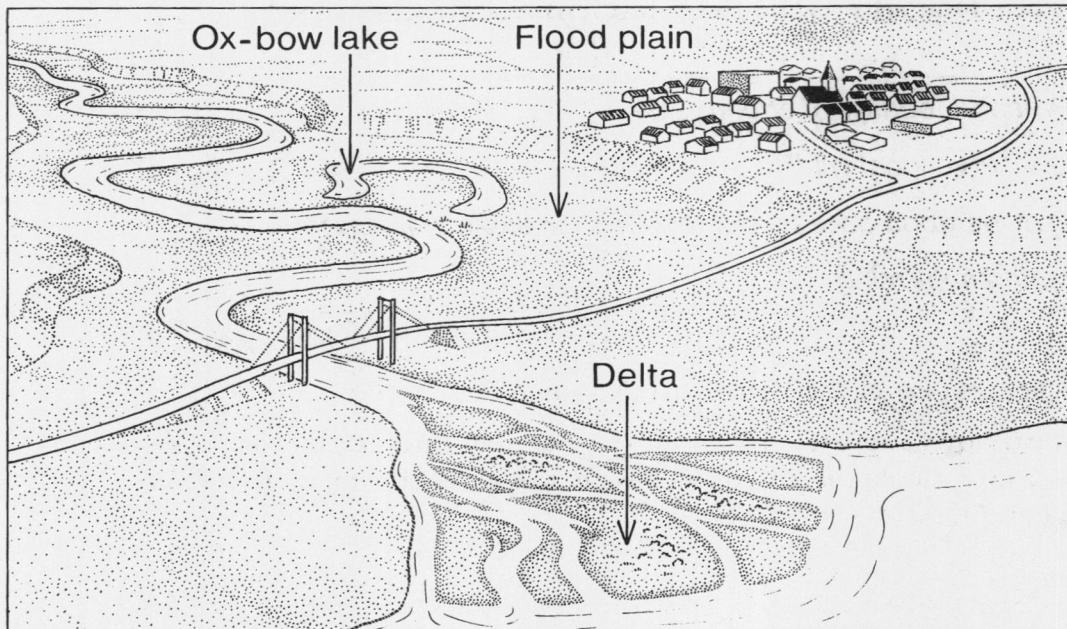

Look at Reference Diagram Q2.

Choose **one** of the named river features shown and **explain** how it was formed.

You may use a diagram(s) to illustrate your answer.

_____

_____

_____

_____

_____

_____

_____

_____

3

Marks

3. **Reference Diagram Q3A:**
   Weather Station Symbol for
   Aberdeen 12 noon, 17th December

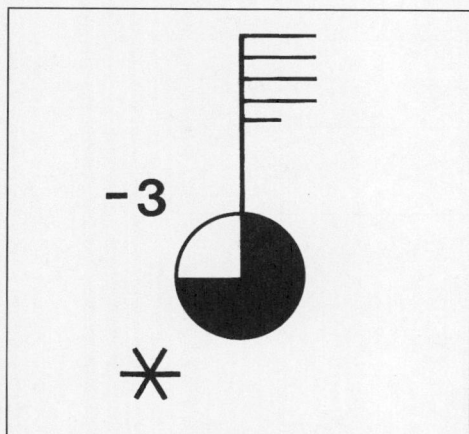

   Reference Diagram Q3B:
   Advertisement for
   Football Match

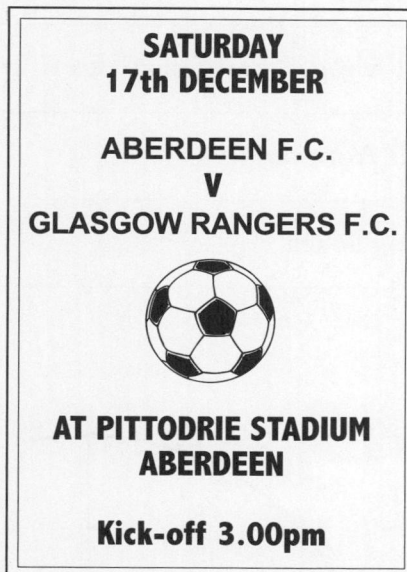

-3

SATURDAY
17th DECEMBER

ABERDEEN F.C.
V
GLASGOW RANGERS F.C.

AT PITTODRIE STADIUM
ABERDEEN

Kick-off 3.00pm

Look at Reference Diagrams Q3A and Q3B.

On Saturday morning the referee decided to postpone this game.

Referring to the weather conditions, give reasons for his decision.

_____

_____

_____

_____

_____

_____

4

[Turn over

Page seven

Marks

4.   **Reference Diagram Q4A: Causes of Sea Pollution**

A	Sewage and industrial waste	40%
B	Wind-blown gases and particles from industry	35%
C	Oil spills from tankers	15%
D	Dumping at sea	10%

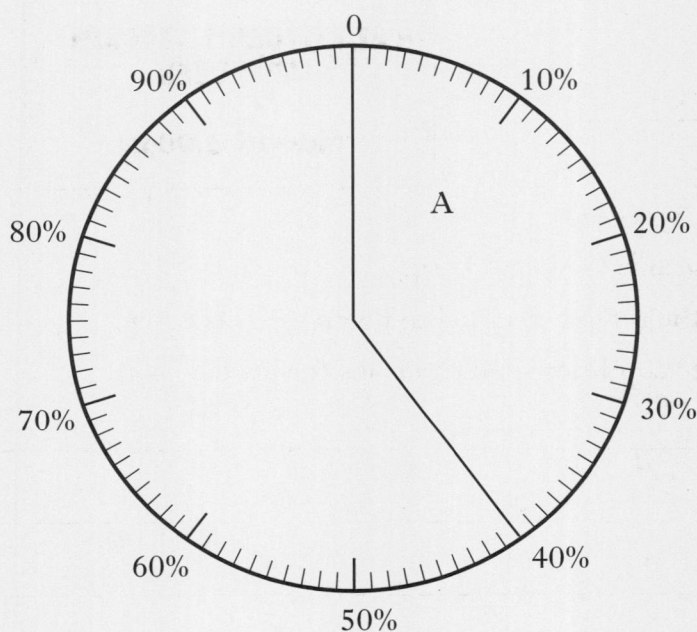

(a)   Complete the pie chart to show the information given in Reference Diagram Q4A.

3

DO NOT WRITE IN THIS MARGIN

KU | ES

Marks

4. (continued)

Reference Diagram Q4B: Methods of reducing Sea Pollution

| **A** | Enforce laws banning dumping at sea |
| **B** | Ensure that sewage is treated before it goes out to sea |

(*b*) Which **one** of the above measures do you think would be the best method of reducing sea pollution?

Choice: **A** or **B** _____

Give reasons for your choice.

_____

_____

_____

_____

_____

_____

3

[Turn over

5.   **Reference Diagram Q5A: Selected Climate Regions**

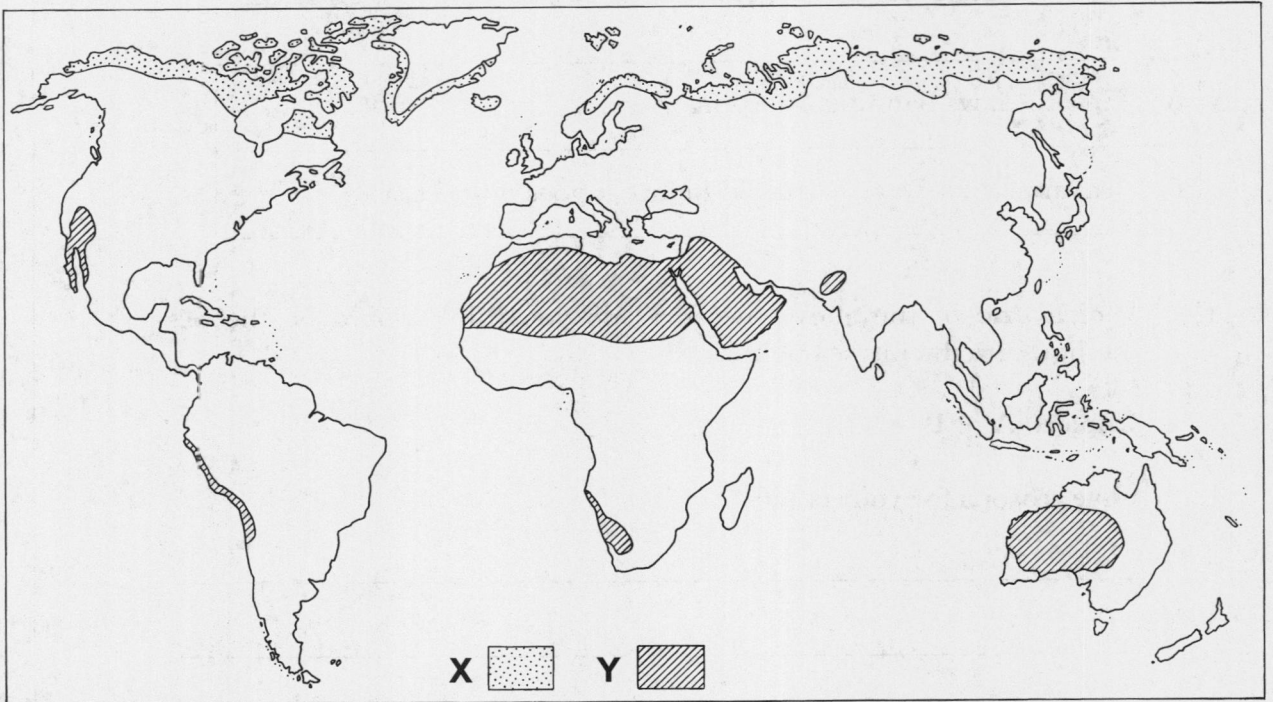

5. (continued)

Reference Diagram Q5B: Climate Graphs for Selected Regions

Region X

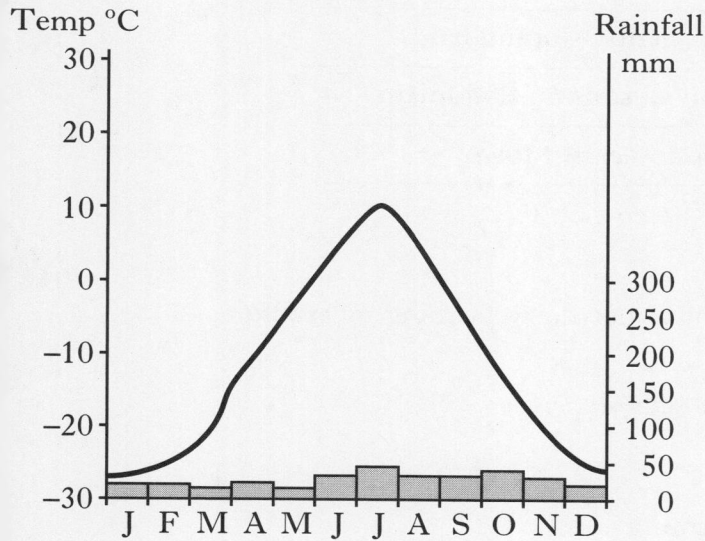

Region Y

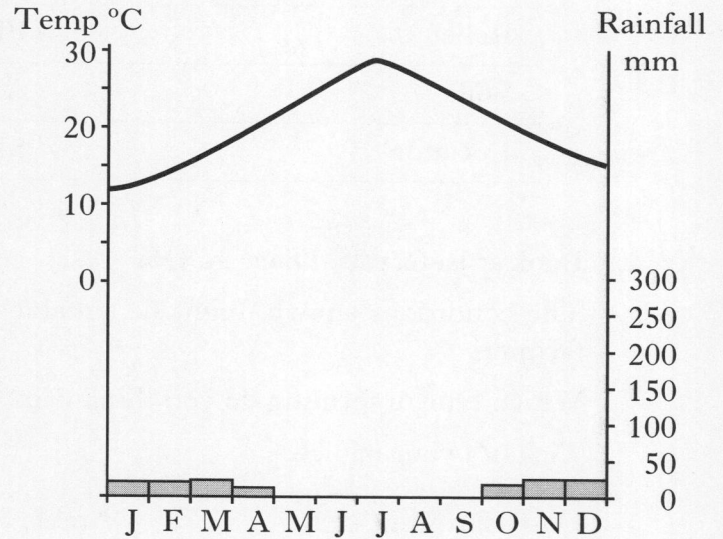

Look at Reference Diagrams Q5A and Q5B.

Both regions X and Y are sparsely populated.

Referring to the climate information, **explain** why it is difficult for people to live and work in each of these regions.

Region X _____

_____

_____

_____

Region Y _____

_____

_____

_____

DO NOT WRITE IN THIS MARGIN

KU | ES

Marks

4

Marks

6. **Reference Diagram Q6: Selected Farm Data**

Average Temperature	January 5° C, July 16° C
Annual Precipitation	600–800 mm
Relief	Flat, gently sloping land
Soils	Alluvial soils on floodplain
Location	5 km to nearest town

Look at Reference Diagram Q6.

The conditions shown might be suitable for either dairy farming or arable farming.

Which type of farming do you think is more likely?

Tick (✓) your choice.

Dairy farming ☐   Arable farming ☐

Give reasons for your choice.

_____

_____

_____

_____

4

Marks

7. **Reference Diagram Q7: A Modern Industrial Landscape**

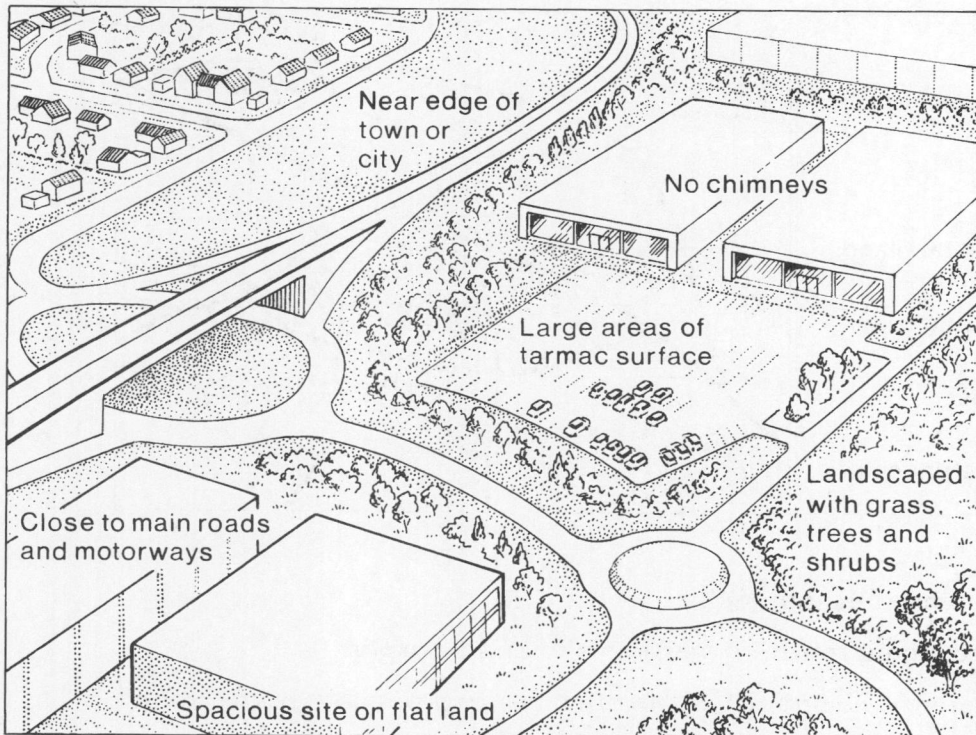

Near edge of town or city

No chimneys

Large areas of tarmac surface

Landscaped with grass, trees and shrubs

Close to main roads and motorways

Spacious site on flat land

Look at Reference Diagram Q7.

Explain why some of the labelled features are typical of a modern industrial landscape.

_____

_____

_____

_____

_____

_____

4

[Turn over

Marks

8. **Reference Diagram Q8: Reducing Traffic Congestion**

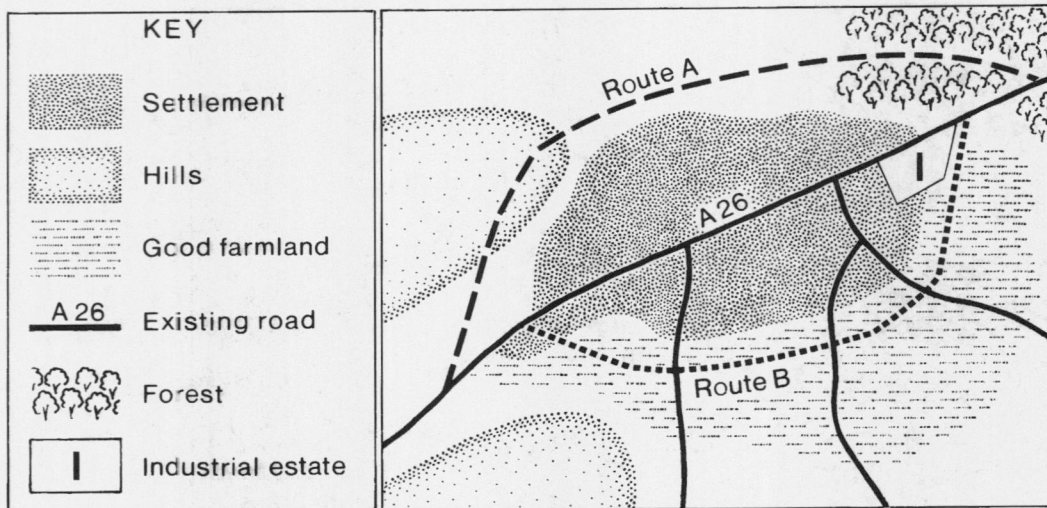

KEY

- Settlement
- Hills
- Good farmland
- A 26 — Existing road
- Forest
- I — Industrial estate

Look at Reference Diagram Q8.

One way of reducing traffic congestion is to build a *bypass*.

Look at Reference Diagram Q8 which shows two possible routes round a settlement.

Which route, A **or** B, do you think would be the better choice?

Tick (✓) your choice.

Route A ☐   Route B ☐

Give reasons for your choice.

_____

_____

_____

_____

_____

_____

4

[Turn over for Question 9 on *Page sixteen*

Marks

9.   **Reference Diagram Q9A: Factors which influence
Death Rates in Europe**

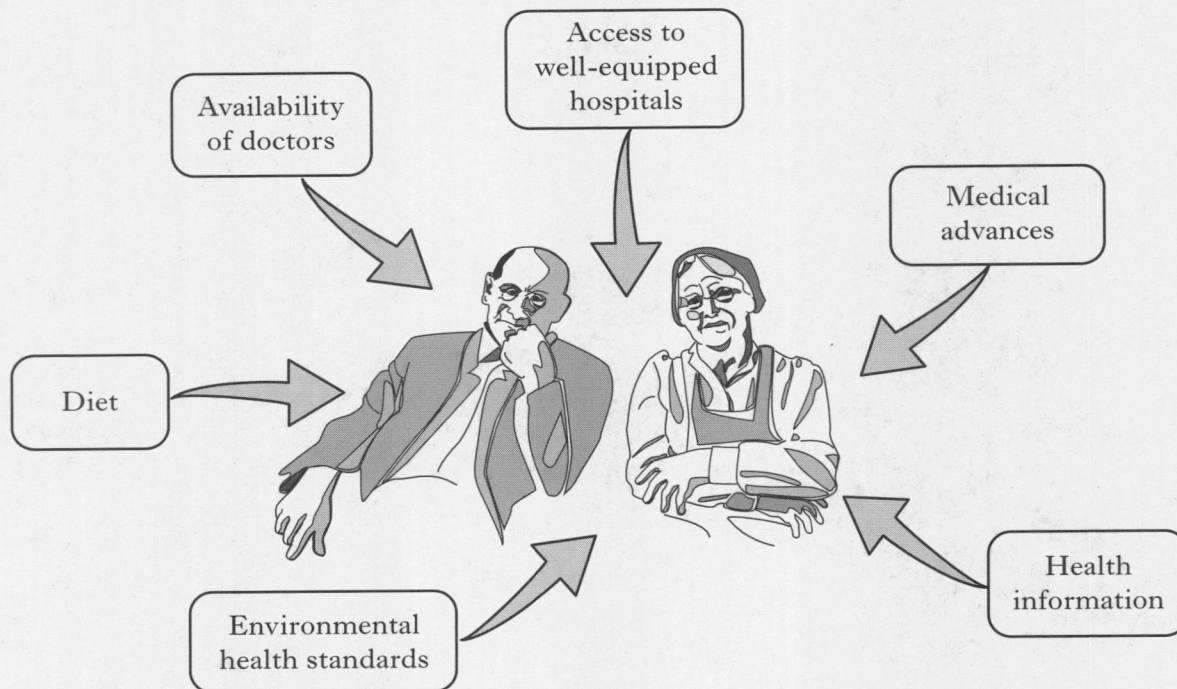

Availability
of doctors

Access to
well-equipped
hospitals

Medical
advances

Diet

Health
information

Environmental
health standards

(*a*)   Study Reference Diagram Q9A.

For any **two** of the factors shown, **explain** how they affect death rates
in Europe.

Factor 1 _____

_____

_____

_____

_____

Factor 2 _____

_____

_____

_____

_____

4

KU | ES

Marks

9. (continued)

Reference Diagram Q9B: Life Expectancy in Italy

Year	Life Expectancy
1950	64
1960	66
1970	70
1980	72
1990	75
2000	78
2005	79

(b) Look at Reference Diagram Q9B.

Name **two** other techniques which could be used to show the information given.

Give reasons for your choices.

Technique 1 _____

Reason(s) _____

_____

_____

_____

Technique 2 _____

Reason(s) _____

_____

_____

_____

4

[Turn over

Marks

10.   **Reference Diagram Q10A: Tied Aid**

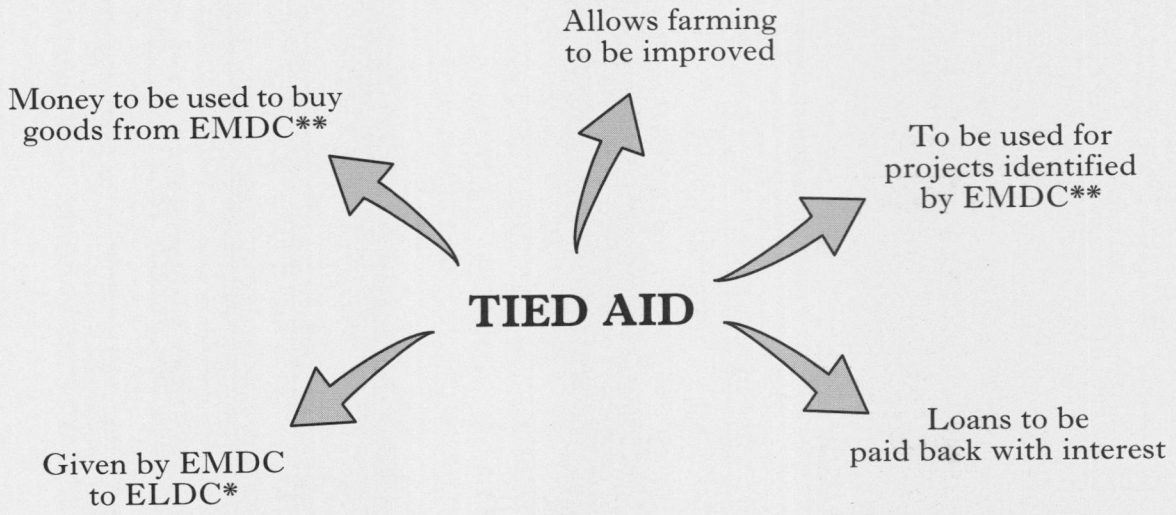

Allows farming
to be improved

Money to be used to buy
goods from EMDC**

To be used for
projects identified
by EMDC**

TIED AID

Given by EMDC
to ELDC*

Loans to be
paid back with interest

    *ELDC = Economically Less Developed Country
    **EMDC = Economically More Developed Country

(a)   Look at Reference Diagram Q10A.

What are the advantages **and** disadvantages of tied aid for ELDCs?

_____

_____

_____

_____

_____

4

Marks

10. (continued)

Reference Diagram Q10B: Education in ELDCs

(b)   Look at Reference Diagram Q10B.

"Education is the best way to improve living conditions in ELDCs."

Do you agree fully with this statement?

Give reasons for your answer.

_____

_____

_____

_____

_____

_____

3

[Turn over for Question 11 on *Page twenty*

Marks

11.   **Reference Diagram Q11: Location of UK and India**

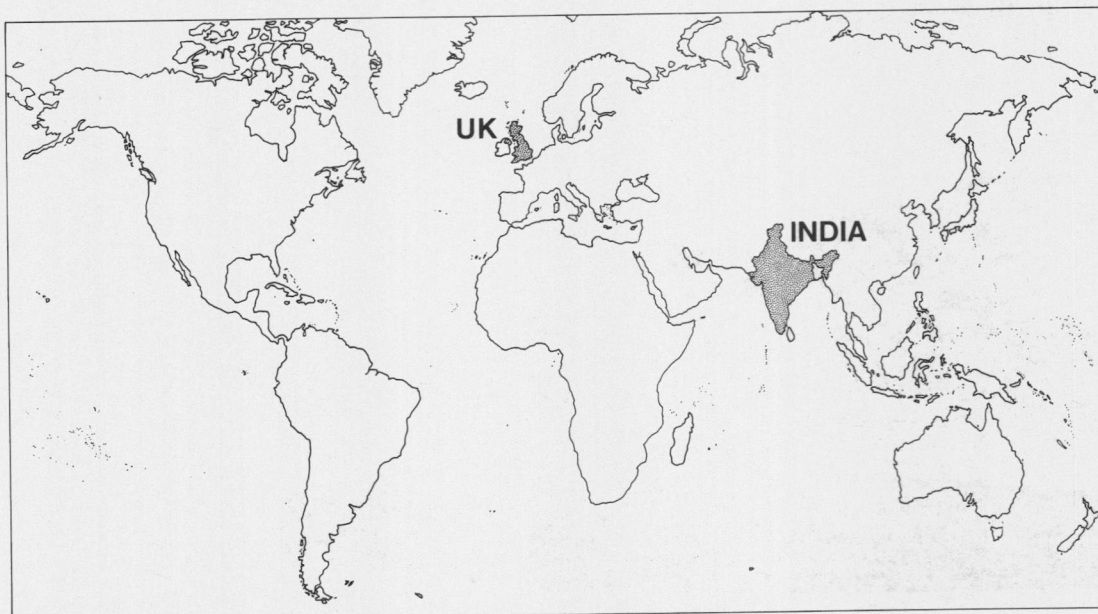

The following statements are about Development and Trade in the UK or India.

Match the letters to the correct country in the table below.

A   Quotas on exports to Germany and France

B   Exports mainly manufactured goods

C   Agriculture employs 66% of population

D   Energy consumption per capita is low

E   Agriculture employs 2% of population

F   High GNP per capita

UK	India

4

[END OF QUESTION PAPER]

[BLANK PAGE]

C

1260/405

NATIONAL
QUALIFICATIONS
2008

FRIDAY, 9 MAY
1.00 PM – 3.00 PM

GEOGRAPHY
STANDARD GRADE
Credit Level

All questions should be attempted.

Candidates should read the questions carefully. Answers should be clearly expressed and relevant.

Credit will always be given for appropriate sketch-maps and diagrams.

Write legibly and neatly, and leave a space of about one centimetre between the lines.

All maps and diagrams in this paper have been printed in black only: no other colours have been used.

SQA

Extract No 1656/104

1:50 000 Scale
Landranger Series

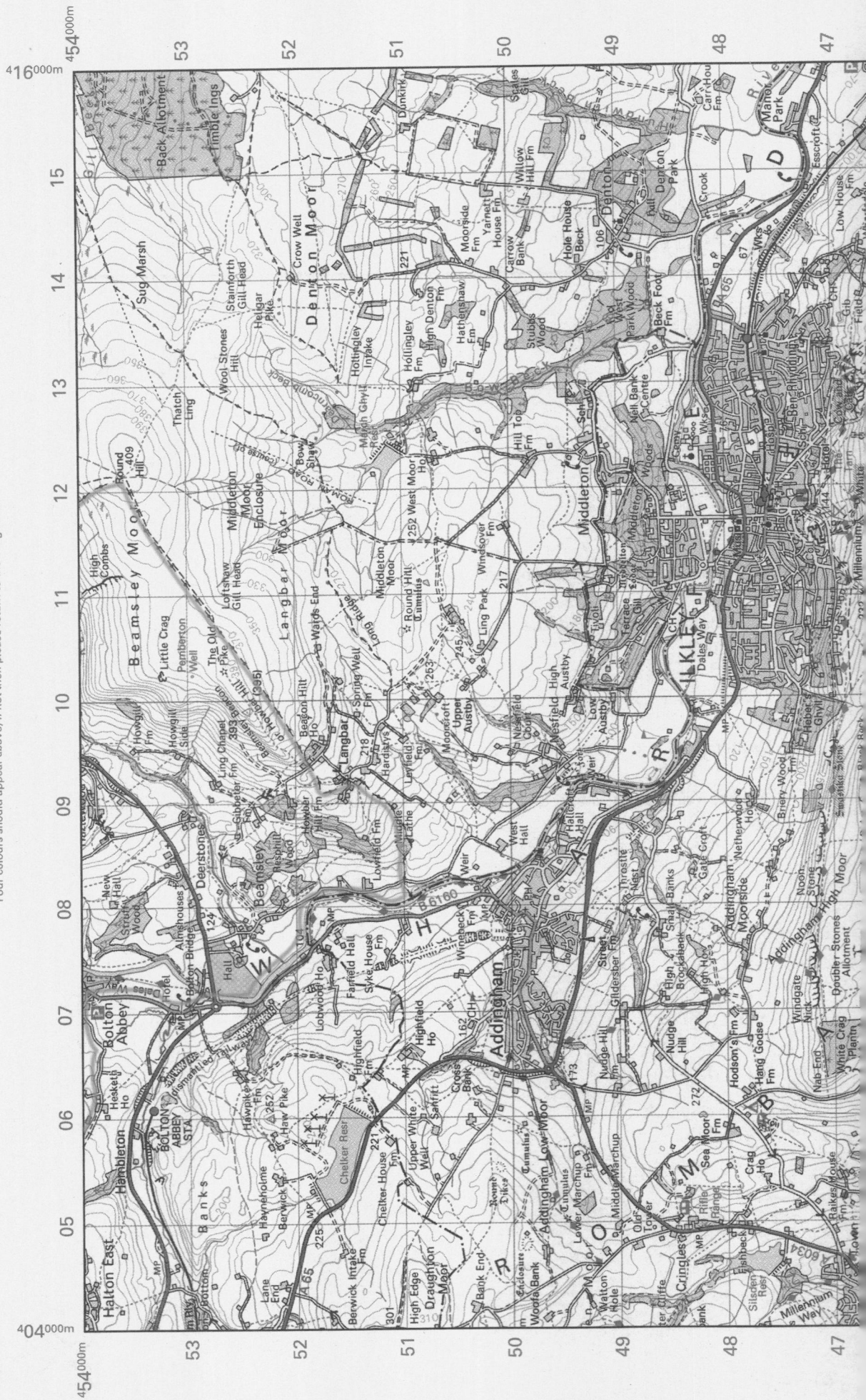

Four colours should appear above; if not then please return to the invigilator.
Four colours should appear above; if not then please return to the invigilator.

Scale 1:50 000

2 centimetres to 1 kilometre (one grid square)

1 mile = 1·6093 kilometres

1 kilometre = 0·6214 mile

Grid North

True North

Magnetic North

Diagrammatic only

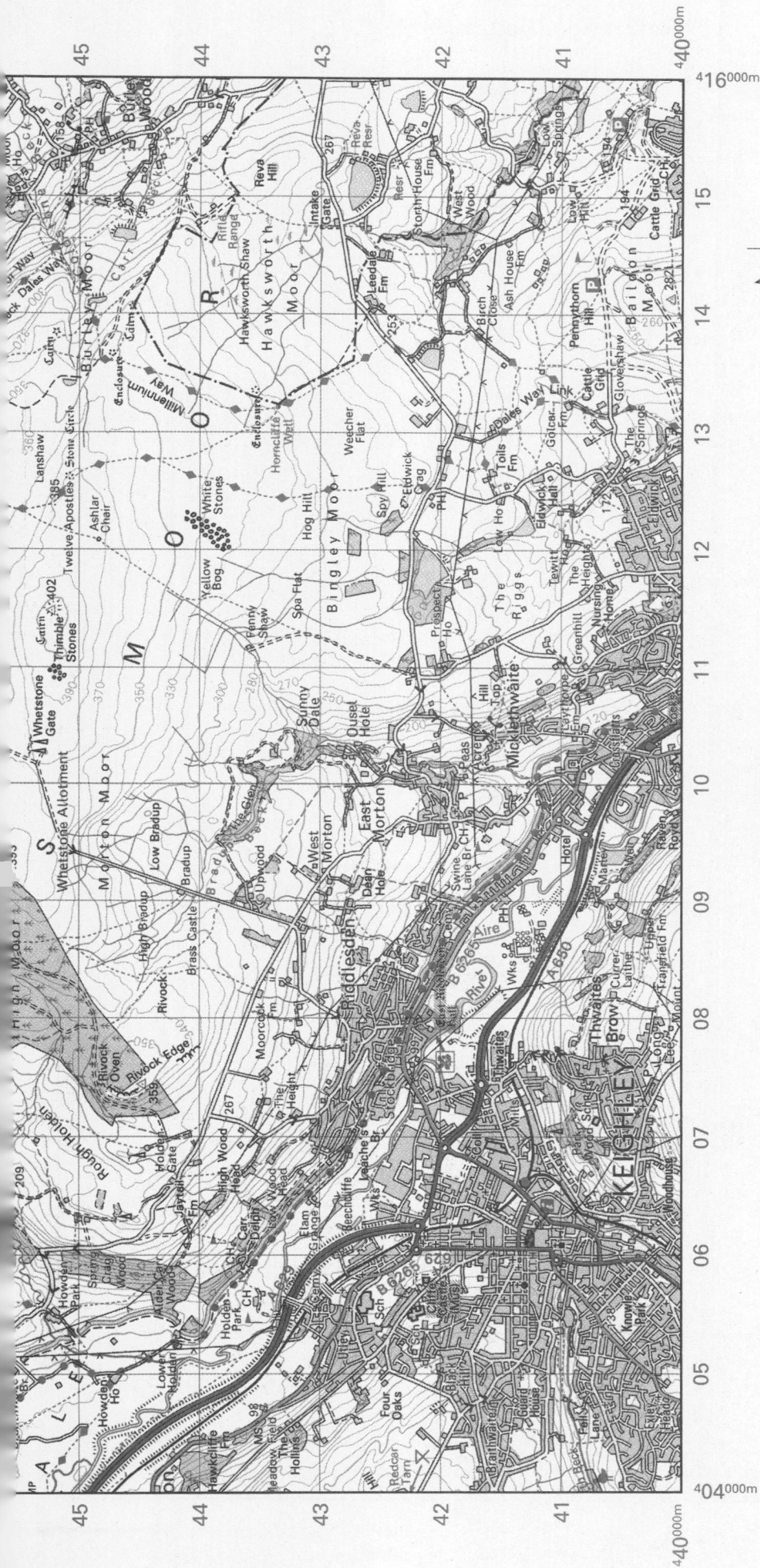

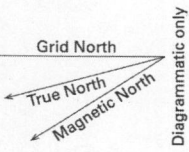

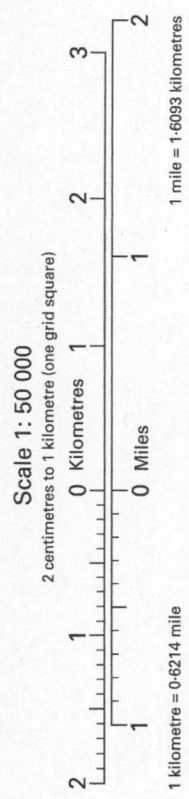

Extract produced by Ordnance Survey 2007. Licence: 100035658
© Crown copyright 2006. All rights reserved.

1.   **Reference Diagram Q1**

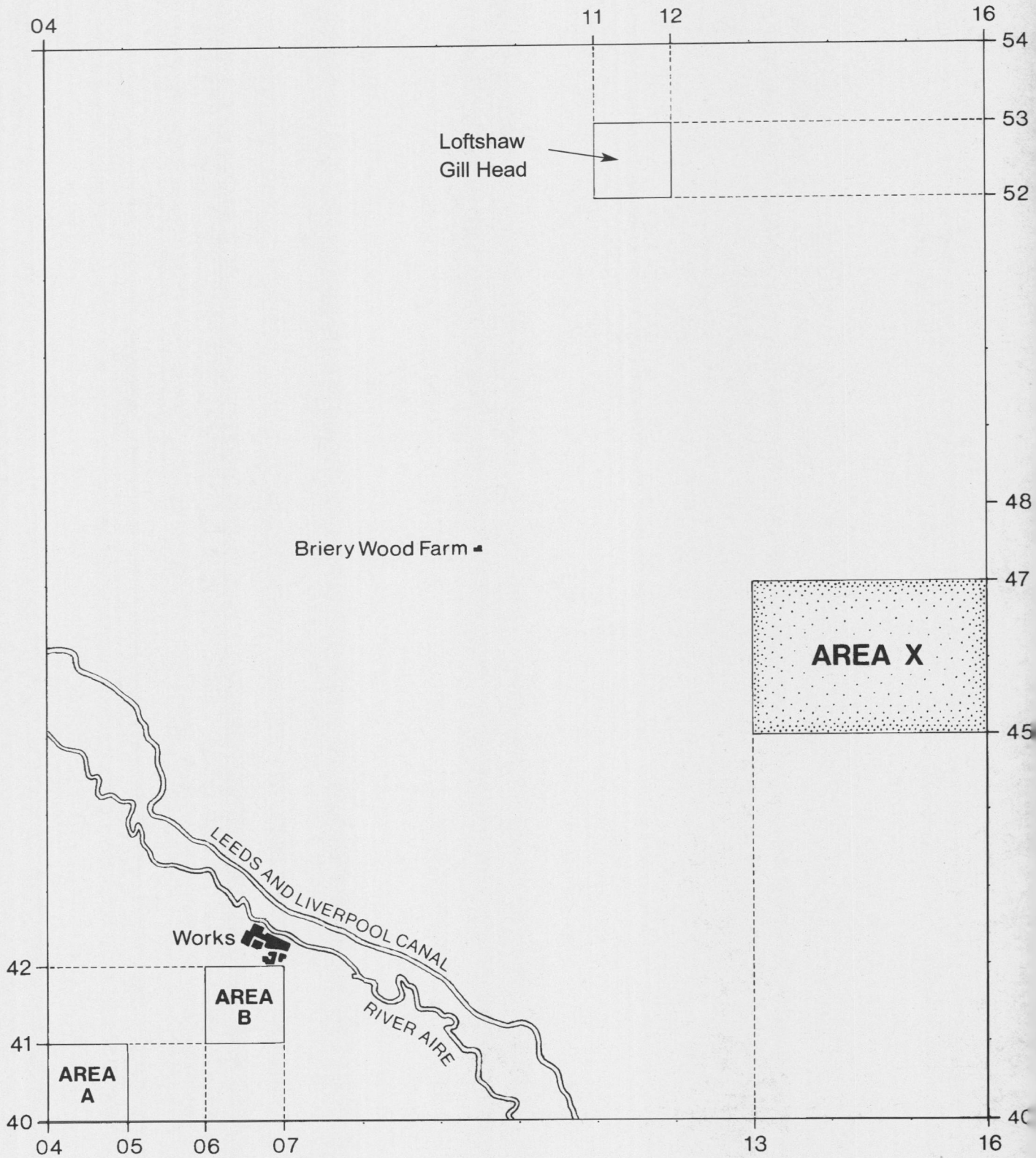

Marks

KU	ES

1. (continued)

This question refers to the OS Map Extract (No 1656/104) of the Ilkley area and Reference Diagram Q1.

(*a*) Describe the **physical** features of the River Aire **and** its valley from 041450 to 099400.

Your answer should **not** refer to the Leeds and Liverpool canal.

KU 4

(*b*) The valley which runs south east from Loftshaw Gill Head in square 1152 is a "V" shaped river valley. Explain how this is likely to have been formed.

You may use diagrams to illustrate your answer.

KU 4

(*c*) Find Briery Wood Farm at 095474.

What are the advantages **and/or** disadvantages of this location for a farm?

ES 4

(*d*) Why might there be conflicts between the various land uses in Area X?

ES 6

(*e*) Describe the differences between the urban environments of Area A (0440) and Area B (0641).

ES 5

(*f*) Find the works in grid square 0642.

Why is this a good site for the works?

ES 5

[Turn over

2.   **Reference Diagram Q2: A Lowland Landscape in Scotland**

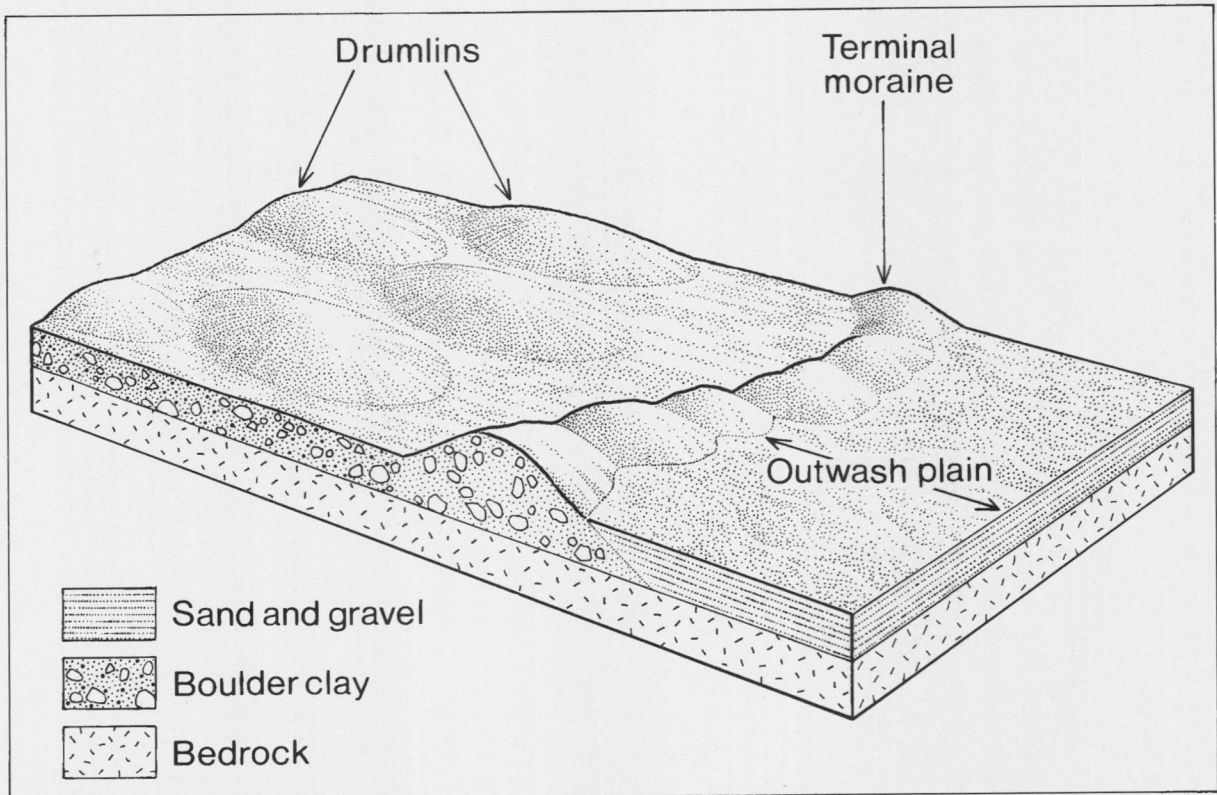

Look at Reference Diagram Q2.

Choose any **two** of the features shown in the diagram and **explain** how they were formed.

You may use diagram(s) to illustrate your answer.

5

3. **Reference Diagram Q3: European Synoptic Chart for noon, 8th July**

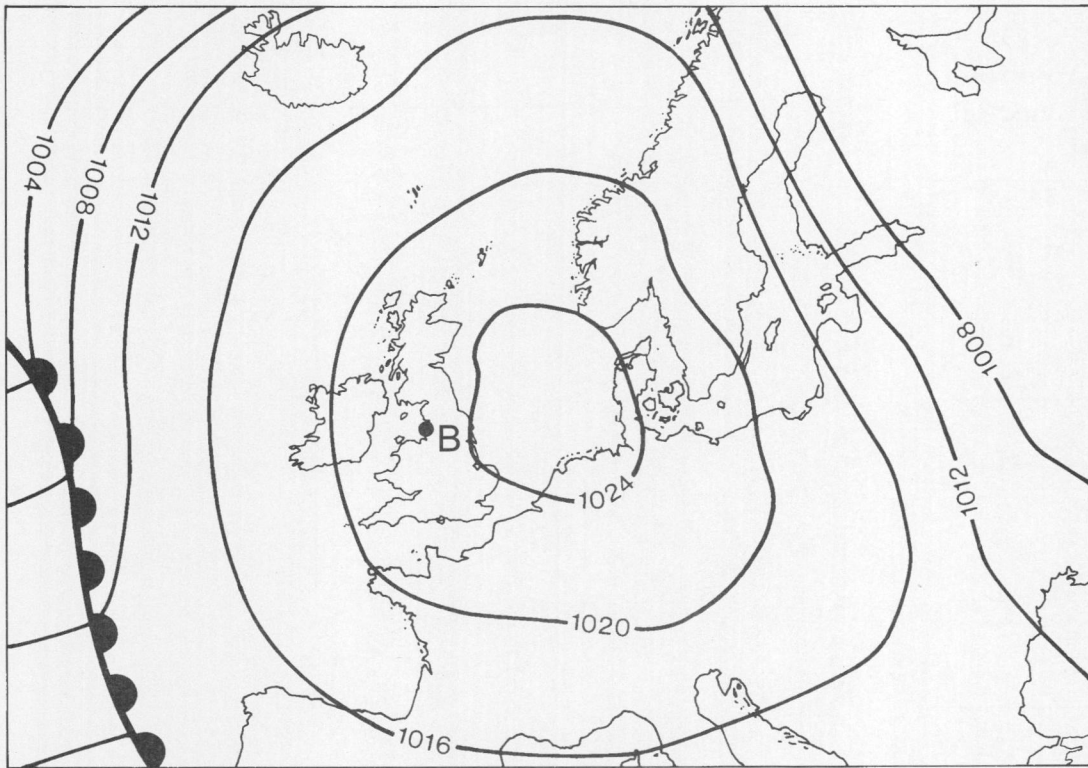

• B – Blackpool

Look at Reference Diagram Q3.

On July 8th Mr McCormack is taking his young family to Blackpool for a one week holiday.

Do you think the weather conditions will be favourable for them?

Give reasons your answer.

5

[Turn over

4.   **Reference Diagram Q4: Brazilian Rainforest Facts**

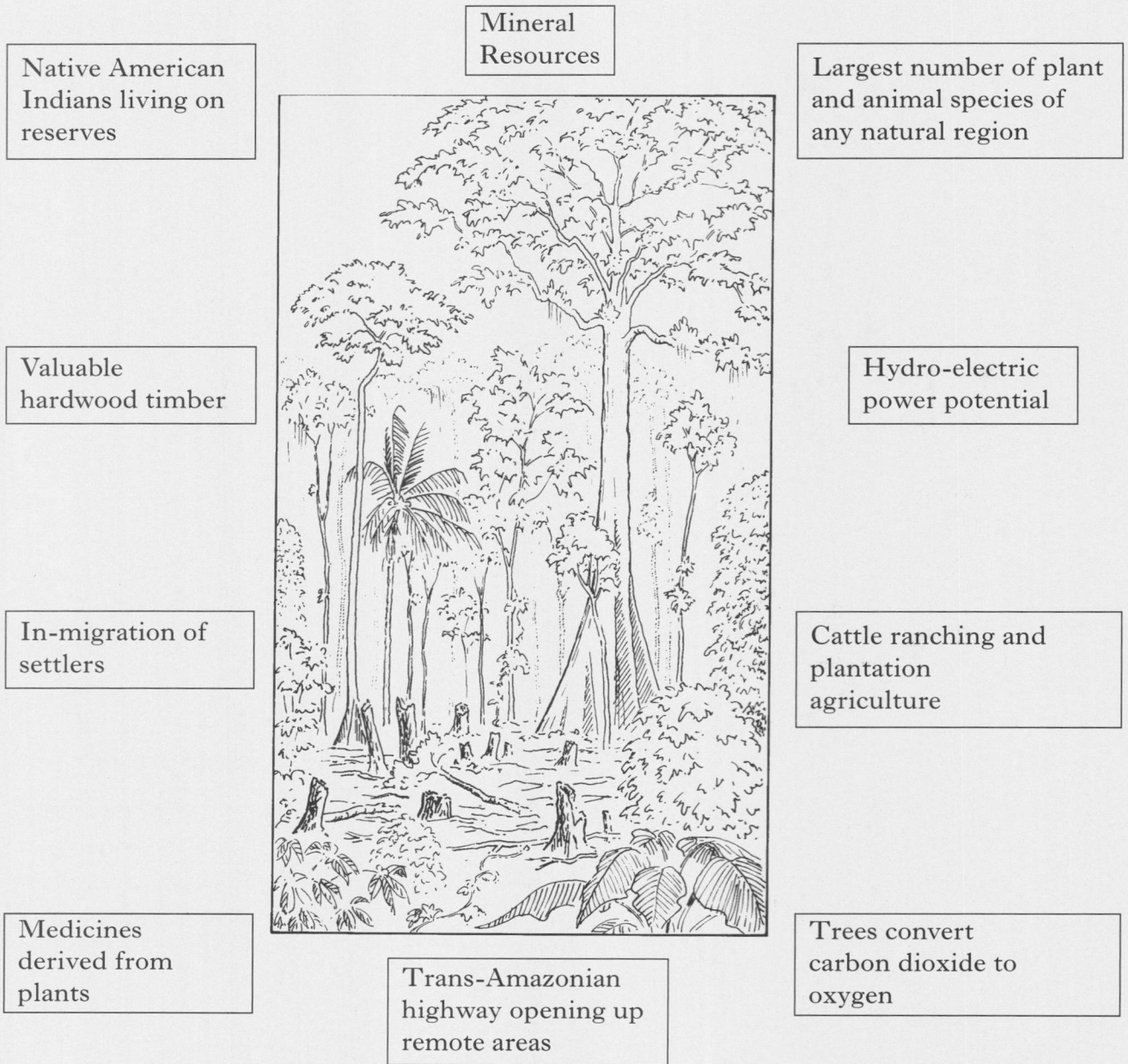

Mineral Resources

Native American Indians living on reserves

Largest number of plant and animal species of any natural region

Valuable hardwood timber

Hydro-electric power potential

In-migration of settlers

Cattle ranching and plantation agriculture

Medicines derived from plants

Trans-Amazonian highway opening up remote areas

Trees convert carbon dioxide to oxygen

Mar

KU

"Cutting down the rainforest in Brazil will affect the whole world more than it will affect Brazil itself."

(Statement by an environment spokesperson)

Look at Reference Diagram Q4 and the statement above.

To what extent do you agree with the statement?

Give reasons for your answer.

5.　　　　**Reference Diagram Q5A:　Location of Housing Types**

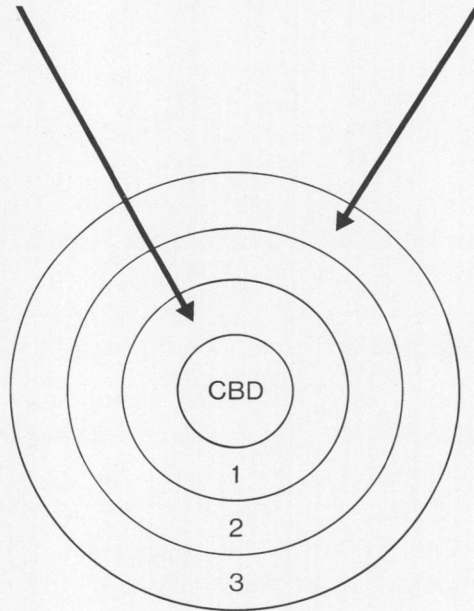

Key to Zones

CBD　Central Business District
1　　19th Century Housing
2　　1930s Inter War Housing
3　　Late 20th Century Housing

	Marks	
	KU	ES

(a)　Look at Reference Diagram Q5A.

　　Explain in detail why there are different types of houses in Zones 1 and 3.　　　　　6

Reference Diagram Q5B:　Statement

"There have been many changes in the CBD."

(b)　Look at Reference Diagram Q5B.

　　What techniques could a group of Geography students use to gather information on changes in the Central Business District (CBD)?

　　Give reasons for your chosen techniques.　　　　　　　　　　　　　　　　5

6. **Reference Diagram Q6A: Eurocentral**

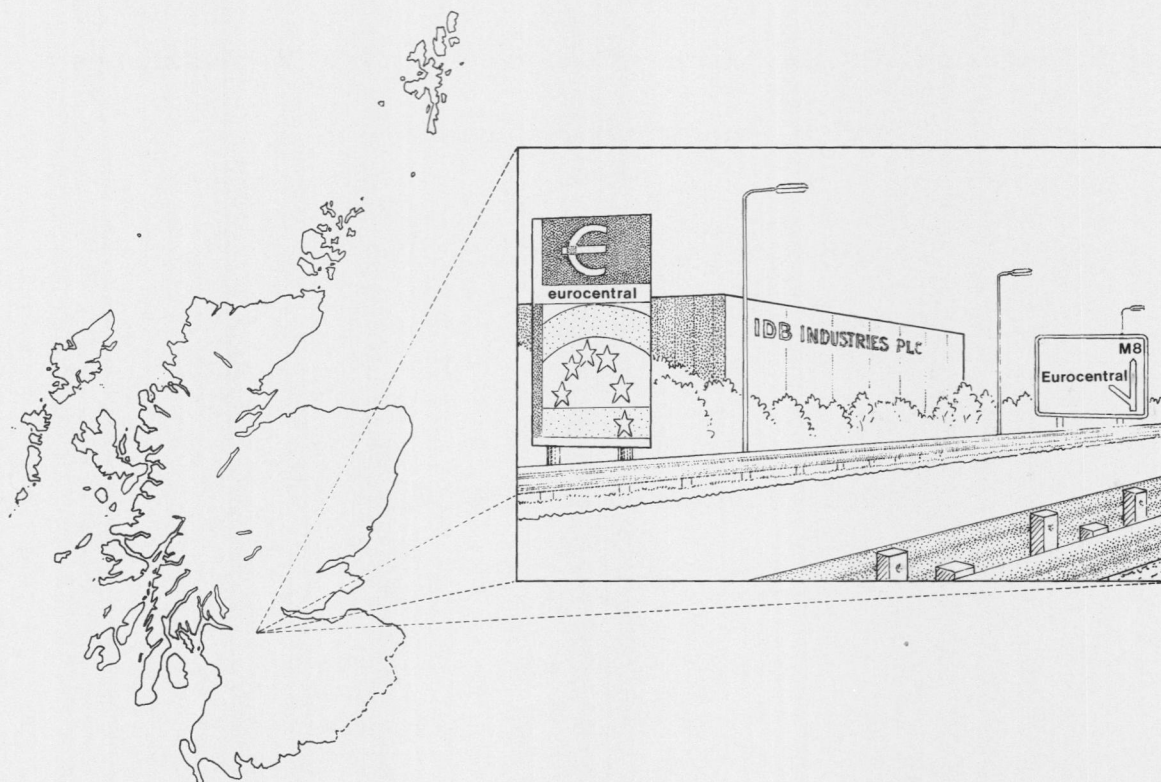

Reference Diagram Q6B: Extract from a News Report

"Eurocentral is a 260 hectare greenfield site where new large-scale industrial development is planned.

This is expected to have a significant impact on the surrounding communities, where old industry has been in decline."

(a) Look at Reference Diagrams Q6A and Q6B.

Describe the benefits **and** problems which new, large-scale industrial development may bring to areas such as this.

6

6. (continued)

Reference Diagram Q6C: North Lanarkshire—Selected Industrial Statistics

Table 1: Employment Categories

Public and other services	30 700
Retailing and wholesale	24 800
Manufacturing	18 700
Finance and business	14 500

Table 2: Unemployment 1996–2002

1996	12 500
1997	10 500
1998	10 000
1999	8500
2000	8000
2001	7500
2002	7000

(*b*) Look at Reference Diagram Q6C.

What other techniques could be used to present the data shown above?

Give reasons for your choices.

5

[Turn over

7.   **Reference Diagram Q7A: Demographic Transition Model**

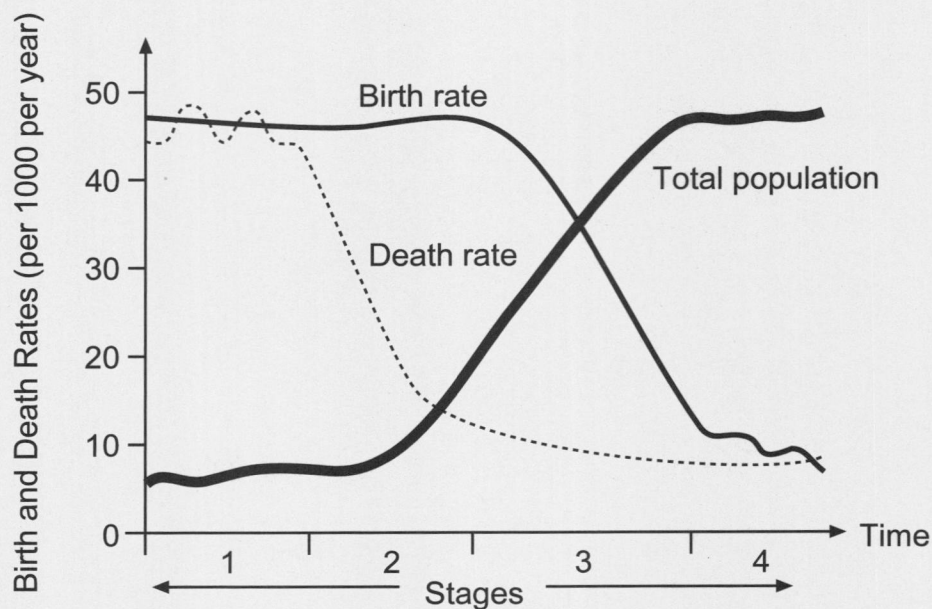

(*a*)   Look at Reference Diagram Q7A.

Describe in detail the changes shown on the Demographic Transition Model from Stage 1 to Stage 4.

Reference Diagram Q7B: Selected Population Data

Country	Crude Birth Rate per 1000	Crude Death Rate per 1000	Natural Increase per 1000
India	23	8	15
Nigeria	38	14	24
UK	10·8	10·1	0·7
USA	14·1	8·3	5·8

(*b*)   Look at Reference Diagram Q7B.

Choose **one** country shown on the table.

Suggest reasons for its **rate of natural increase**.

4

Marks
KU | ES

8.

Reference Diagram Q8A: Japan's Exports

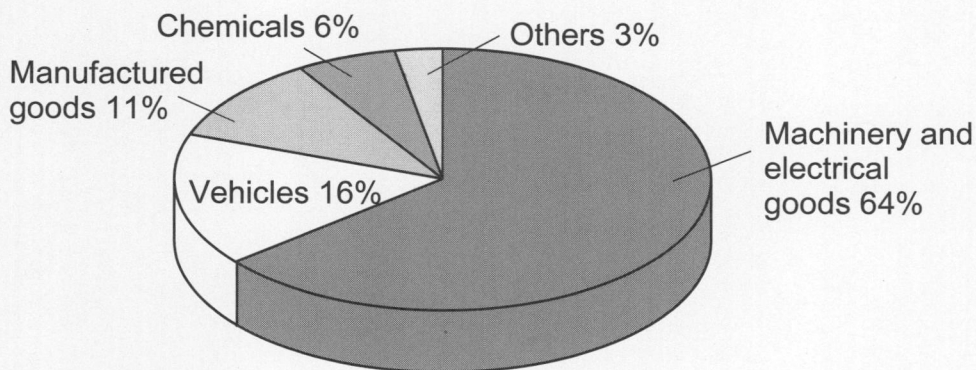

Chemicals 6%

Others 3%

Manufactured goods 11%

Machinery and electrical goods 64%

Vehicles 16%

Reference Diagram Q8B: Japan's Imports

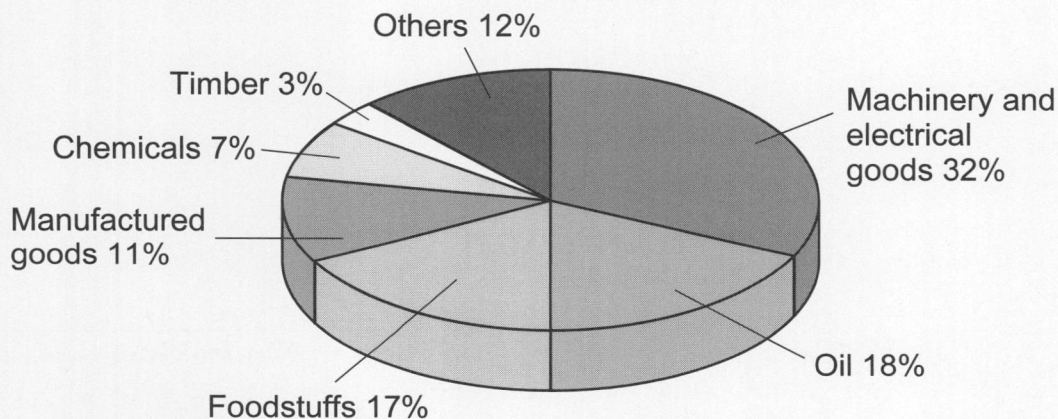

Others 12%

Timber 3%

Chemicals 7%

Machinery and electrical goods 32%

Manufactured goods 11%

Oil 18%

Foodstuffs 17%

Look at Reference Diagrams Q8A and Q8B.

(a) What are the main differences between Japan's exports and imports? | | 3

(b) Japan has a large trade surplus and is the world's second biggest trading nation.

Explain why Japan depends on world trade for the success of its economy. | 3 |

[END OF QUESTION PAPER]

[BLANK PAGE]

[BLANK PAGE]

[BLANK PAGE]

Official SQA Past Papers: Credit Geography 2008

[BLANK PAGE]

Acknowledgements

Leckie and Leckie is grateful to the copyright holders, as credited, for permission to use their material:
This product includes mapping data reproduced by permission of Ordnance Survey on behalf of HMSO. © Crown Copyright 2008. All rights reserved. Ordnance Survey Licence number 100036009.

The following companies have very generously given permission to reproduce their copyright material free of charge:
Toyota PLC for a photograph (2004 Credit paper p 9).